BSA A50 & A65 Twins Owners Workshop Manual

by Mark Reynolds

Models covered from 1962 to 1973

499 cc	A50	Star Twin
499 cc	A50C	Cyclone
499 cc	A50IR	Royal Star
499 cc	A50W	Wasp
654 cc	A65	Star Twin
654 cc	A65IT	Thunderbolt
654 cc	A652H	Hornet
654 cc	A65FB	Firebird
654 cc	A65L	Lightning
654 cc	A65R	Rocket
654 cc	A652S	Spitfire

ISBN 978 0 85696 155 7

© Haynes Group Limited 1990

Printed in India   *(155-2P6)*

ABCDE
FGHIJ
KLMNO
PQRS
3

Haynes Group Limited
Sparkford, Yeovil,
Somerset BA22 7JJ, England

Haynes North America, Inc
2801 Townsgate Road,
Suite 340, Thousand Oaks,
CA 91361, USA

Acknowledgements

Our thanks are due to BSA Motor Cycles Limited for their assistance. Brian Horsfall gave the necessary assistance with the overhaul and devised the ingenious methods for overcoming the lack of service tools. Les Brazier arranged and took the photographs that accompany the test. Jeff Clew advised about the presentation of, and edited the text. We would also like to acknowledge the help of the Avon Rubber Company, who kindly supplied the illustrations relating to tyre fitting and Amal Limited for their carburettor illustrations.

Thanks are also due to Ken Doyle of Ken Doyle motorcycles, Sherborne who provided the A65 Spitfire MK II featured on the front cover.

About this manual

The author of this manual has the conviction that the only way in which a meaningful and easy-to-follow text can be written is to carry out the work himself, under conditions similar to those found in the average household. As a result the hands seen in the photographs are those of the author. Even the machines are not new; examples which have covered a considerable mileage are selected, so that the conditions encountered would be typical of those encountered by the average rider/owner. Unless specially mentioned, and therefore considered essential, BSA service tools have not been used. There are invariably alternative means of slackening or remove some vital component when service tools are not available, but risk of damage is to be avoided at all costs. Each of the Chapters is divided into numbered sections. Within the sections are numbered paragraphs. Cross-reference throughout the manual is quite straightforward and logical. For example, when reference is made 'see Section 6.2' it means Section 6, paragraph 2 in the same chapter. If another chapter were meant, the reference would read 'See Chapter 2, section 6.2'. All photographs are captioned with a section/paragraph number to which they refer, and are always relevant to the chapter text adjacent.

Figure numbers (usually line illustrations) appear in numerical order within a given Chapter. Fig.1.1 therefore refers to the first figure in Chapter 1. Left hand and right descriptions of the machines and their component parts refer to the left and right when the rider is seated facing forward.

Motorcycle manufacturers continually make changes to specifications and recommendations, and these, when notified, are incorporated at the earliest opportunity.

Whilst every care is taken to ensure that the information in this manual is correct no liability can be accepted by the authors or publishers for loss, damage or injury caused by any errors in or omissions from the information given.

Introduction to the
BSA 500/650 unit construction vertical twins

Many will be surprised to learn that the development of a vertical twin BSA motor cylce commenced so far back as August 1939, although the outbreak of war only a month later curtailed all further work for several years. It was not until November 1945 that a proto-type was available for the press to ride and even then there was an embargo on the release of any details. The announcement of BSA's first vertical twin was made in September 1946. This was a 495 cc model and quite a few of these twins appeared in the following years Clubmans T.T. This first model, the A7. was refined through the years and in 1950 at the Earls Court Show a 646 cc model was released. During the next twelve years the 500 cc and 650 cc. Models were further refined; The old fashioned plunger rear suspension being discarded in favour of the swinging arm. In addition to frame improvements many engine modifications were made culminating with the inauguration of the unit consturruction twins in 1962. The range of motorcylces that BSA produced from 1962 onwards was very comprehensive. It included fast road machines in the Lightning and Thunderbolt, and off road machines in the form of the Firebird Scrambler and the track racer, the Hornet. Similarly with the 500 cc models there was an option for a road or 'track' machine.

In 1970 the 500 cc models and the dated styling of the BSA range were discarded. A new frame and forks of greater rigidity and lighter weight has been designed, together with conical wheel hubs, with their breaking action moulded on car practice. Indicators, following Japanese practice, were then fitted and it is their specifications that applied until all production ceased in 1972.

Contents

NB.   Specifications and general descriptions are given in each Chapter immediately after the list of contents.
Where applicable Fault diagnosis is given at the end of the appropriate Chapter.

Side view of A65 650 Lightning

Ordering spare parts

When ordering spare parts for any of the BSA unit construction twins, it is advisable to deal direct with an official BSA agent who will be able to supply many of the items ex stock. Parts cannot be obtained direct from the BSA manufacturers; all orders must be routed through an approved agent, even if the parts required are not held in stock.

Always quote the engine and frame numbers in full. Include any letters before or after the number itself. The frame number will be found stamped on the left hand front down-tube adjacent to the steering head. The engine number is stamped on the left hand crankcase, immediately below the base of the cylinder barrel.

Use only parts of genuine BSA manufacture. Pattern parts are available but in many instances they will have an adverse effect on performance and/or reliability. Some complete units are available on a service exchange basis affording an economic method of repair without having to wait for parts to be reconditioned. Details of the parts available, which include, petrol tanks, front forks, front and rear frames, clutch plates, brake shoes etc., can be obtained from any BSA agent. It follows that the parts to be exchanged must be acceptable before factory reconditioned replacements can be supplied.

Some of the more expendable parts such as spark plugs, bulbs, tyres, oils and greases etc., can be obtained from accessory shops and motor factors, who have convenient opening hours, charge lower prices and can often be found not far from home. It is also possible to obtain parts on a Mail Order basis from a number of specialists who advertise regularly in the motor cycle magazines.

Location of engine number

Location of frame number

Routine maintenance

Periodic routine maintenance is a continuous process that commences immediately the machine is used. It must be carried out at specified mileage recordings or on a calendar date basis if the machine is not used regularly - whichever falls soonest. Maintenance should be regarded as an insurance policy rather than a chore, because it will help keep the machine in peak condition and ensure long, trouble-free service. It has the additional benefit of giving early warning of any faults that may develop and will act as a regular safety check , to the obvious benefit of both rider and machine alike.

The various maintenance tasks are described under their respective mileage and calendar date headings. Accompanying diagrams have been added, where necessary. It should be remembered that the interval between the various maintenance tasks serves only as a guide. As the machine get older or is used under particularly arduous conditions, it would be advisable to reduce the period between each check.

Some of the tasks are described in detail, where they are not mentioned fully as a routine maintenance item in the text. If a specific item is mentioned but not described in detail, it will be covered fully in the appropriate Chapter. No special tools are required for the normal routine maintenance tasks. The tools contained in the kit supplied with every new machine will prove adequate for each task but if they are not available, the tools found in the average household should suffice.

Weekly, or every 200 miles

Check oil tank contents and replenish if necessary. Do not overfill. Oil brake pedal pivot and cable, also all other exposed control cables and joints. Check battery acid level, chain adjustments and tyre pressures.

Monthly, or every 1,000 miles

Complete the maintenance tasks listed under the preceding weekly heading, then the following additional items:

Check the gearbox oil level as follows: Ensure that the machine is standing absolutely upright on its wheels on level ground. On 1962 to 1967 models, unscrew the oil level screw from the gearbox drain plug; a small amount of oil will drain out of the stand pipe. Unscrew the filler cap from the crankcase right-hand half and slowly add oil of the correct type until **fresh** oil can be seen dripping from the stand pipe. When the surplus oil has drained off, refit the level and filler plugs. On 1968 and later models, unscrew the filler plug/dipstick, wipe it clean and screw it in again; the oil level should be up to the mark on the dipstick. Add or remove oil as necessary, then refit the dipstick. Grease the swing arm pivot and the speedometer drive. Grease the wheel hubs and brake cam spindles, where grease nipples are fitted. Note that both these points should be greased sparingly, to prevent grease from entering the brake drums. Drain and refill the primary chaincase. Check tightness of all bolts.

Two monthly, or every 2,000 miles

Complete all the checks listed under the weekly and monthly headings, then the following items:

Drain the oil tank whilst the oil is warm and clean out the filters. Refill with fresh oil of the correct viscosity. Remove, clean and lubricate the final drive chain. Dismantle and clean the carburettor(s); clean and re-oil the air filter element (if fitted). Inject a few drops of oil through the lubricator on the left hand end of the dynamo. Check the tappets, the contact breaker and sparking plug gaps.

Six monthly, or every 5,000 miles

Again, complete all the routine maintenance tasks listed previously, then the following additional tasks:

Drain the gearbox oil whilst it is warm and refill with fresh oil of the correct viscosity. Drain also the telescopic fork legs and refill each leg with the correct amount of new oil. Check play in the wheel bearings, renew and replace if play is evident.

Yearly, or every 10,000 miles

After completing the weekly, two monthly and six monthly tasks, continue with the following additional items:

Decarbonise the engine and give a complete top overhaul. Repack the wheel bearings and the headrace bearings with grease. Check the bushes, bearings and oil seals of the front forks, replacing any that require attention.

Note:

No specific mention has been made of the lighting system, horn and speedometer, each of which must be in correct working order to satisfy statutory requirements. It is assumed these items will receive frequent attention, especially since a fault in any one will immediately be obvious to the rider of the machine. This also applies to the tyres, which must be renewed when they have reached the prescribed limit of wear or if they show any other form of defect that may place the safety of the machine and/or rider at risk.

Remember there is no stage at any point in the lift of the machine when a routine maintenance task can be ignored, or safety checks neglected.

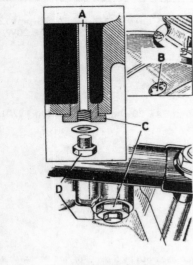

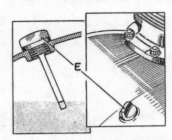

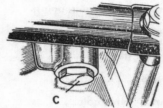

A Stand pipe
B Filler cap
C Drain plug
D Level screw
E Filler plug/dipstick

Checking the gearbox oil level

1962 to 1967 models

1968 on models

RM1. Engine oil must be changed every two months

RM2. Check the primary chaincase oil content regularly (see Fig. 1.9)

RM3. Check the gearbox oil content regularly and change the oil every six months

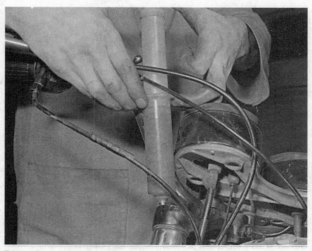

RM4. Drain and refill front forks every 6 months

Quick glance maintenance data

Engine	Capacity: 6.0 pints/3.4 litres (1962), 5.5 pints/3.1 litres (1963–1965), 5.0 litres (1966 on). Grade: Summer SAE 40 or 50, Winter SAE 20 or 30, or SAE 20W/50 engine oil
Primary chaincase	1/4 pint (140 cc) of SAE 10W/30 or 10W/40 or SAE 20 engine oil
Gearbox	7/8 pint (500 cc) of hypoid gear oil, SAE 90 (EP90)
Front forks – per leg	1/3 pint (190 cc) of SAE 10W/30 or 10W/40 or SAE 20 engine oil (1962 to 1970), TQF (1971 on)
Bearings and other greasing points	Multi-purpose high melting-point lithium based grease
Final drive chain	Commercial chain lubricant
Contact breaker gap	0.015 in (0.38 mm)
Spark plug gap	0.020 – 0.025 in (0.51 – 0.64 mm)
Tyre pressures – tyres cold	21 psi front, 22 psi rear

Note: Pressures given are basic setting for low speed with a rider of 154 lb. For high speed increase both by 5 psi – for heavier loads add 1 psi (front) or 2 psi (rear) for every extra 28 lb load

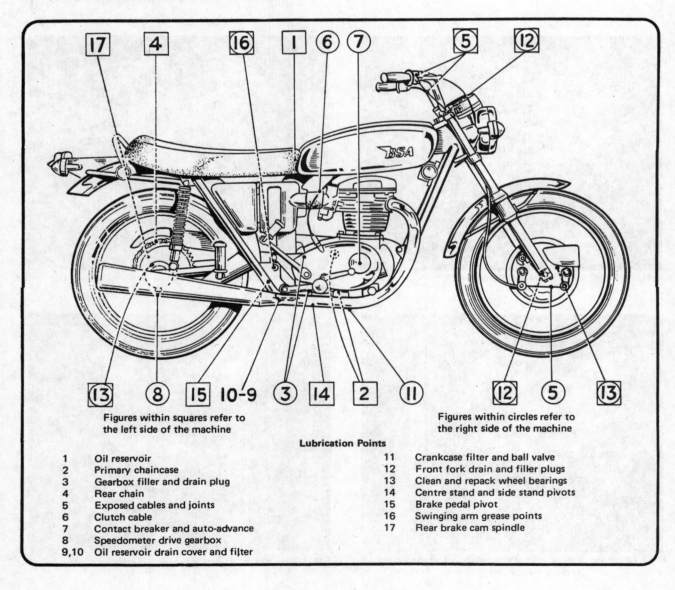

Figures within squares refer to the left side of the machine

Figures within circles refer to the right side of the machine

Lubrication Points

1	Oil reservoir	11	Crankcase filter and ball valve
2	Primary chaincase	12	Front fork drain and filler plugs
3	Gearbox filler and drain plug	13	Clean and repack wheel bearings
4	Rear chain	14	Centre stand and side stand pivots
5	Exposed cables and joints	15	Brake pedal pivot
6	Clutch cable	16	Swinging arm grease points
7	Contact breaker and auto-advance	17	Rear brake cam spindle
8	Speedometer drive gearbox		
9,10	Oil reservoir drain cover and filter		

Chapter 1 Engine, clutch and gearbox

Contents

Specifications

The BSA 650/500 cc unit construction vertical twins all employ the same basic engine/gear unit in which the gearbox is an integral part of the engine assembly. The same dismantling and reassembly procedure is applicable to all the 650/500 cc unit construction models. Unless otherwise stated, engine specifications listed below apply to 500 and 650 cc models.

Engine: (Lightning, Thunderbolt, Spitfire, Firebird, Hornet, Royal Star)

Type ...	Twin cylinder, vertically mounted
Cylinder head ...	Aluminium alloy
Cylinder barrel ...	Cast iron
Bore ...	75 mm* 65.5+
Stroke ...	74 mm
Capacity ...	654 cc* 499 cc+
Compression ratio ...	9.1
	10,2 (Spitfire only)

* All 650 cc models
+ 500 cc models

Crankshaft:

Main bearing left hand (drive side) ...	Single lipped roller bearing
Size ...	2.812 in. x 1.125 in. x 0.812 in.
Main bearing right hand (timing side) ...	Phosphor bronze bush
Size ...	1.624 in x 1.5 in. x 0.94 in.

Big end journal diameter	1.6865in. — 1.687 in.
Minimum regrind diameter	1.6565 in. — 1.657 in.
Permissible end float	0.0015 in. — 0.003 in.

Connecting rods:

Length (centres)	6.0 in.
Big end bearings - type	Vandervall V.P.D2
Bearing side clearance	.024 in.

Camshafts:

Journal bearing diameter - left hand	0.810 in. — 0.8105 in.
- right hand	0.8735 in. — 874 in.
End float	Nil (spring loaded)
Cam lift - Inlet and exhaust	0.306 in.
Base circle diameter	0.812 in.

Tappets:

Clearance (cold) - Inlet	0.008 in.
- Exhaust	0.010 in.

Pistons:

Clearance (top of skirt)	0.0045 in. — 0.0050 in.* 0.0062 in. — 0.0065 in.
(bottom of skirt)	0.0019 in. — 0.0024 in.* 0.0032 in. — 0.0035 in.
Oversizes available	+ 10 thou. + 20 thou. + 40 thou.

Piston rings:

Compression rings (two rings, tapered)

Width	0.0615 in. — 0.0625 in.* 0.114 in. — 0.121 in.
Radial depth	0.098 in. — 0.104 in.* 0.0615 in. — 0.0625 in.
End gap	0.007 in. — 0.012 in.* 0.008 in. — 0.013 in.

Oil control ring

Radial depth	0.098 in. — 0.104 in.* 0.114 in. — 0.121 in.
Width	0.124 in. — .125 in.* 0.124 in. — 0.125 in.
End gap	0.008 in. — 0.013 in.

* 500 cc models

Valves:

Stem diameter - Inlet	0.3095 in. — 0.310 in.
- Exhaust	0.309 in. — 0.3095 in.
Head diameter - Inlet	1.450 in. — 1.455 in.* 1.595 in. — 1.60 in.
- Exhaust	1.312 in. — 1.317 in.* 1.407 in. — 1.412 in.

Valve guides:

Material	Cast iron (high grade)
Bore diameter - Inlet and exhaust	0.312 in. — 0.313 in.
Outside diameter - Inlet and exhaust	0.5005 in. — 0.501 in.
Length - Inlet	1.96 in. — 1.97 in.
- Exhaust	1.96 in. — 1.97 in.

Valve springs:

Free length - Outer	2 1/32 in. (* 1 ¾ 1968 onwards)
- Inner	1 5/8 in. (1 7/16 in. 1968 onwards)

* 500 cc models

Torque wrench settings: (Dry)

	ft lb
Flywheel bolts	30
Connecting rod bolts	22
Cylinder head bolts (3/8 in.)	25
Cylinder head bolts (5/16 in.)	25
Cylinder head nuts (3/8 in.)	26
Cylinder barrel nuts (5/16 in.)	18
Oil pump stud nuts	7
Clutch centre nut	70 - 75
Kickstarter ratchet nut	60
Rotor fixing nut	60
Stator fixing nuts	10 - 15
Crankshaft pinion nut	60
Manifold stud nuts (5/16 in.)	12.5
Manifold stud nuts (1/4 in.)	6
Carburettor flange nuts	10
Zener diode fixing nut	17

Valve timing:

Inlet valve opens	...	...	...	...	...	...	...	51º BTDC*	40 BTDC+
Inlet valve closes	...	...	...	...	...	...	...	68º ABDC*	60 ABDC+
Exhaust valve opens	...	...	...	...	...	...	...	78º BBDC*	65º BBDC+
Exhaust valve closes	...	...	...	...	...	...	...	37º ATDC*	35º ATDC+

Note: All valve clearances must be set at 0.015 in. whilst checking.

* Sports cam (650 cc models only)
+ Standard cam

1 General description

The engine fitted to the BSA unit construction twins of the combined engine and gearbox type in which the gearbox casting forms an integral part with the right hand or timing side crankcase; all the castings being of aluminium alloy. The cylinder head has cast in austernitic valve seat inserts and houses the overhead valves which are actuated by rocker arms enclosed within the detachable rocker box. The push rods are of aluminium alloy with hardened end pieces.

H section connecting rods of Hinduminium alloy with detachable caps and steel backed shell bearings carry aluminium alloy die-cast pistons each with two compression and one oil scraper ring. The two throw crankshaft has a detachable shrunk-on flywheel which is retained by three high tensile steel bolts. The cast iron barrel houses the press fit tappet block.

The BSA twins have a combined inlet and exhaust camshaft which runs in sintered bronze bearings at the upper rear of the engine crankcase. The camshaft is driven by a system of timing gears on the right hand end of the crankshaft. The drive side or left hand end of the camshaft operates the rotary breather whilst the gear type oil pump is driven off the timing side mainshaft and the twin points with their automatic advance retard unit are driven off the idler gearwheel. Power from the engine is transmitted in the conventional manner through the engine sprocket and primary chain to the multiplate clutch unit which embodies a shock absorber; thence through the gearbox and to the rear wheel.

2 Operations with engine in frame

It is not necessary to remove the engine unit from the frame unless the crankshaft assembly and/or main bearings require attention. Most operations can be accomplished with the engine in situ, such as:
1 Removal and replacement of cylinder head.
2 Removal and replacement of cylinder barrel and pistons.
3 Removal and replacement of alternator.
4 Removal and replacement of primary drive or gearbox components.
5 Removal and replacement of oil pump and contact breaker assembly.

If a major overhaul is envisaged it is better to remove the engine at the outset. If a bench vice is available and this can be adapted to hold the engine work is made much easier!

3 Operations with engine removed

1 Removal and replacement of main bearings.
2 Removal and replacement of crankshaft components.
3 Removal and replacement of camshafts.
4 Removal and replacement of timing gear bushes.

4 Method of engine/gearbox removal

As described previously the engine and gearbox are built as a unit, although it is not necessary to remove the unit complete in order to gain access to either component unless a major overhaul is contemplated. It is not possible to dismantle the engine fully until the engine/gearbox unit has been removed from the frame and refitting cannot take place until the crankcases have been reassembled. The crankcases cannot be separated unless both the outer and inner gearbox cones are removed, exposing the gear clusters.

5 Removing the engine/gearbox unit. Removal of outer timing side cover

It must be stressed at this point that before any work is done on the engine it is advisable to be well armed with both a comprehensive socket set and an impact screwdriver to tackle often burred crosshead screws. In preparing this manual methods have been devised to obviate the use of factory tools, which can be both expensive and difficult to obtain. With the basic requirements of a good tool kit and some initiative most if not all problems (e.g. extraction of gear wheels, clutch centre body) can be handled easily.
1 Place the machine on the centre stand and make sure it is standing firmly on level ground. If it has a steering damper tighten this down firmly.
2 Drain the oil tank by removing its drain plug and catch the oil in a suitable receptacle. Drain the crankcase by removing the plate secured by four bolts beneath the main engine casting. Drain the gearbox. Its drain plug is situated underneath the gearbox casting at the rear. Drain the primary chaincase by removing the cross head screw at the lowest point on the primary chaincase.

With the engine totally drained work can be carried out in a clean and methodical manner.
3 Turn off the petrol taps and disconnect the petrol pipes by unscrewing the petrol tank unions.
4 Remove the tank by unscrewing the central tank nut which is beneath the rubber 'grommet' on top of the tank. The tank can now be lifted clear of the machine.
5 Unscrew the tap(s) of the carburettor body(s) and remove the slide(s) complete. Hang these component(s) well clear of the engine so they do not get in the way of further work.
6 Detach the carburettor body(s) complete by removing the retaining nuts and washers from the inlet manifold. Place them away from the working area. If air filters are fitted it will be necessary to remove these first.
7 Remove spark plug leads and ignition coils by undoing their clamp nuts. When removing the ignition coils take care to remove the associated wires at their snap connector points.
8 Remove the rocker feed oil pipe by unscrewing the union at the rear of the rocker box.
9 Remove the exhaust pipes by slackening off the clamp rings at the exhaust manifolds, and removing the bolts securing the exhaust pipes and silencers to the frame brackets. The exhaust pipes and silencers can then be pulled clear of the machine. The exhaust pipes may need to be tapped clear of the exhaust ports with a hide mallet.

Models fitted with a balance pipe between both exhaust pipe bends should have their respective clamps slackened first. Remove the complete exhaust system.
10 Remove the tachometer cable (if fitted) from the front of the timing case and the speedometer cable from below the gearbox.
11 Remove the footrests. On early models the footrests are

mounted on tapers, each being secured with a nut and washer. Note, however, that although the right-hand footrest nut has a normal right-hand thread, the left-hand footrest nut has a left-hand thread. 'Oil-in-frame' frames (1971 to 1973) have footrests secured by nuts and bolts. After removing the nuts give the footrests a sharp tap with a hide mallet and they will come away.

12 Remove the brake rod from the rear brake pedal. Rotate the brake pedal so that it will not get in the way of engine removal.

13 Remove the kickstarter and gear levers from the timing case.

14 Undo the three peripheral screws of the timing case cover and remove. Remove the timing case cover complete with the points cover plate intact.

15 The clutch arm is now exposed at the rear of the timing side. Remove the cable first from handlebar lever and then slip nipple out of the clutch arm and remove cable from the timing side crankcase.

16 Remove the two wires from the contact breaker points by undoing the small nuts retaining them. Carefully remove these wires from the vicinity of the engine. They can be pulled through a grommet just above the clutch arm.

17 Beneath the engine and at the rear, break the alternator leads at their respective snap connectors.

18 Remove rear chain guard from its frame brackets.

19 Locate the split link of the rear chain and remove with the aid of a pair of pliers. This operation is made easier if the link is positioned on the rear wheel sprocket . Remove the chain.

20 Pull off the feed and return oil pipes from their metal counter parts behind the gearbox and to the right hand side of the rear engine plates.

21 Remove the cylinder head steady plate.

22 Remove the horn and the front engine mounting bolt. The horn can be disconnected from the wiring harness at its snap connectors.

23 Remove the lower central engine mounting bolt which passes through the frame tubes and a lug on the lowest part of the engine crankcase.

24 Remove the engine plates and the four retaining bolts at the rear of the engine/gearbox. The engine is now in a position to be lifted clear of the frame free of obstructions.

25 Remove the engine from the frame by tilting it forward so lifting that the rear end is raised upwards, then twist the unit and lift out from the left hand side.

The BSA engine is a heavy unit and it is advisable to have a second person ready to lend a hand. Lifting the engine from the machine is made easier if one person stands astride the machine, bending down to grasp the engine.

5.4 Tank mounting bolt is recessed:

5.4a Centre bolt fixing is withdrawn

5.5 Remove carburettor top and slide complete

5.10 Tachometer drive located forward of the timing check

5.10a Speedometer drive is located beneath the gearbox

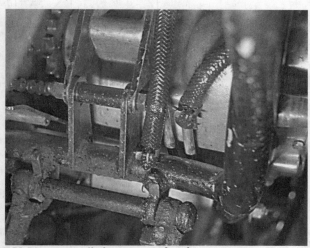

5.20 Disconnect oil pipes at rear of engine

5.25 Lift the engine out sideways

6 Dismantling the engine

1   Before connecting work on the engine unit the external surfaces should be cleaned thoroughly with a rag moistened with paraffin or a proprietary de-greaser to remove road grit and grime. Cleanliness when working on an engine is essential if good results from the rebuild is to be achieved.
2   Never use undo force to remove any stubborn part unless mention is made of this requirement. There is invariably good reason why a part is difficult to remove often because the dismantling operation has been tackled in the wrong sequence.
    If an engine stand is available or the engine can be mounted in a vice, as mentioned before, work will be made much easier.

7 Dismantling the engine - removing the cylinder head, barrel and pistons

1   Remove the rocker box by taking off the six nuts and washers, break the joint by a light tap with a mallet and remove the cover.
2   To gain access to the front two cylinder head bolts the exhaust valve rocker assembly must be removed. Loosen the exhaust tappets and withdraw the two pushrods. Unscrew the nut retaining the exhaust rocker shaft at the front right hand side and tap the shaft out to the left with a blunt punch. Note the position of the spring and thrust washers and remove exhaust rockers.
3   Slacken the inlet tappet and remove the pushrods.

4   Remove the cylinder head by first slackening off the four long centre bolts then the short one in the push rod tunnel and finally the four nuts, unscrewing each a little at a time to avoid distorting the head. Remove the five bolts and four nuts, tap the head with a mallet to break its seal and lift off.
5   Remove the cylinder barrel by unscrewing the eight cylinder base nuts, once again tap the barrel to break the cylinder base seal and lift off carefully. If possible have a friend at hand to steady the pistons and place a clean rag in the crankcase mouth, so that no dirt or grit can enter the engine.
6   Roatate the engine so that the pistons are at the top of their stroke and with the aid of a small screwdriver or some pointed instrument extract the circlips of both pistons and tap out the gudgeon pins. It is useful here to have assistance to steady the pistons. Mark the pistons and connecting rods so that on reassembly the left hand piston goes back on the lefthand rod, etc.
    Scribe 'EX' on the inside of each piston skirt so that they go back together facing the right way. If this procedure is not carried out, especially if high compression pistons are used, the valves may hit the piston when the engine is running.
    At this stage in the proceedings if the crankshaft and its associated components are to be inspected the crankcases must be split. This however entails dismantling the primary drive and clutch together with the gearbox and the removal of the inner timing side cover which reveals the gearbox internals and the camwheel with its idler gear and the gear type oil pump.

8 Dismantling the primary drive, clutch and removal of alternator and gearbox sprocket

1   Take out the twelve screws securing the primary chain case outer cover, two of these have aluminium washers and are the level and drain screws. Tap the chain case to break the seal and remove. If the chaincase has not previously been drained of oil place a receptacle underneath it, to catch any oil.
2   Remove the alternator stator by taking off the three self locking nuts securing it. Take care nto to damage the coils when pulling clear of the three retaining studs.
3   To remove the rotor place a bar through the small ends, resting on the crankcase mouths, to lock the crankshaft and

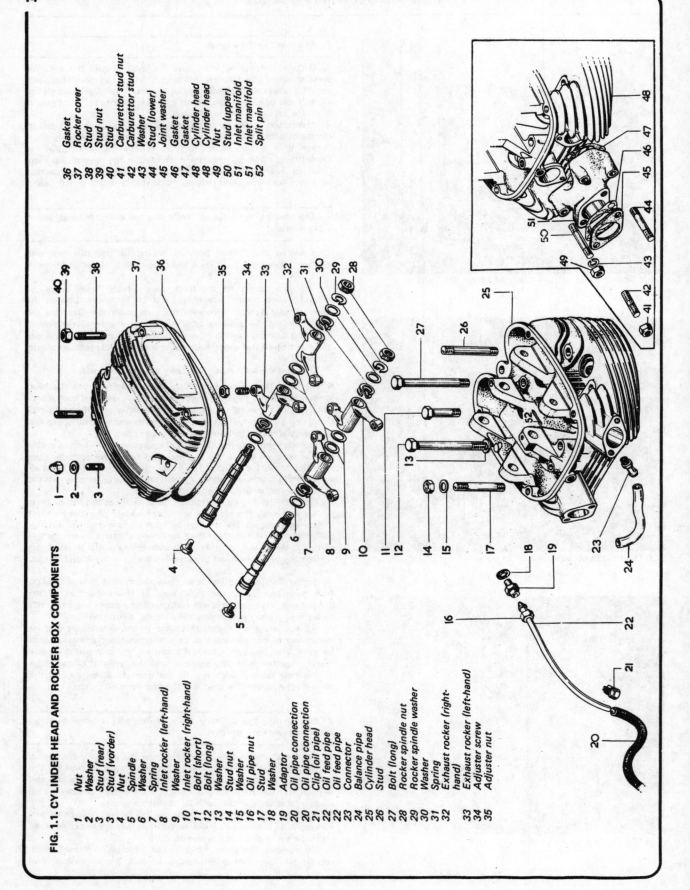

FIG. 1.1. CYLINDER HEAD AND ROCKER BOX COMPONENTS

1 Nut
2 Washer
3 Stud (rear)
3 Stud (vorder)
4 Nut
5 Spindle
6 Washer
7 Spring
8 Inlet rocker (left-hand)
9 Washer
10 Inlet rocker (right-hand)
11 Bolt (short)
12 Bolt (long)
13 Washer
14 Stud nut
15 Washer
16 Oil pipe nut
17 Stud
18 Washer
19 Adaptor
20 Oil pipe connection
20 Oil pipe connection
21 Clip (oil pipe)
22 Oil feed pipe
22 Oil feed pipe
23 Connector
24 Balance pipe
25 Cylinder head
26 Stud
27 Bolt (long)
28 Rocker spindle nut
29 Rocker spindle washer
30 Washer
31 Spring
32 Exhaust rocker (right-hand)
33 Exhaust rocker (left-hand)
34 Adjuster screw
35 Adjuster nut
36 Gasket
37 Rocker cover
38 Stud
39 Stud nut
40 Stud
41 Carburettor stud nut
42 Carburettor stud
43 Washer
44 Stud (lower)
45 Joint washer
46 Gasket
47 Gasket
48 Cylinder head
48 Cylinder head
49 Nut
50 Stud (upper)
51 Inlet manifold
51 Inlet manifold
52 Split pin

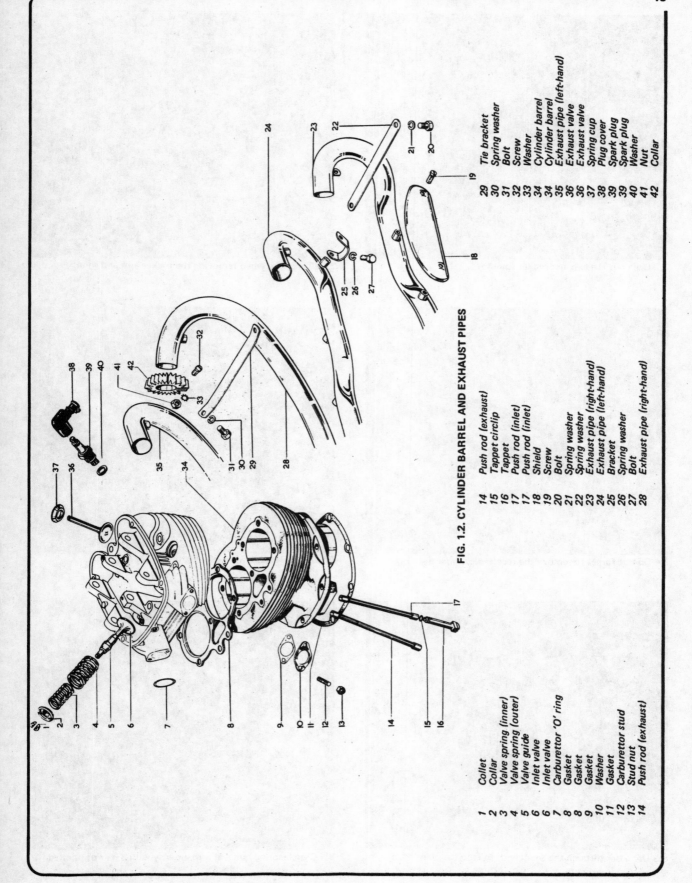

FIG. 1.2. CYLINDER BARREL AND EXHAUST PIPES

1 Collet
2 Collar
3 Valve spring (inner)
4 Valve spring (outer)
5 Valve guide
6 Inlet valve
6 Inlet valve
7 Carburettor 'O' ring
8 Gasket
8 Gasket
9 Gasket
10 Washer
11 Gasket
12 Carburettor stud
13 Stud nut
14 Push rod (exhaust)

14 Push rod (exhaust)
15 Tappet circlip
16 Tappet
17 Push rod (inlet)
17 Push rod (inlet)
18 Shield
19 Screw
20 Bolt
21 Spring washer
22 Spring washer
23 Exhaust pipe (right-hand)
24 Exhaust pipe (left-hand)
25 Bracket
26 Spring washer
27 Bolt
28 Exhaust pipe (right-hand)

29 Tie bracket
30 Spring washer
31 Bolt
32 Screw
33 Washer
34 Cylinder barrel
34 Cylinder barrel
35 Exhaust pipe (left-hand)
36 Exhaust valve
36 Exhaust valve
37 Spring cup
38 Plug cover
39 Spark plug
39 Spark plug
40 Washer
41 Nut
42 Collar

7.2 Drift out the exhaust rocker spindle

7.2a Take care when removing the rocker arm and shims

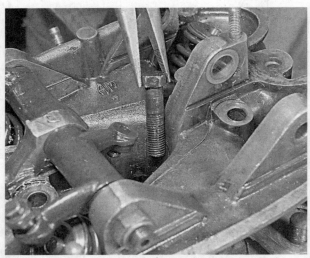

7.4 Do not forget to remove the recessed cylinder head bolt

7.5 Do not let cam followers fall into the engine

8.3 Use a bar through the small ends to stop the crankshaft turning

8.4 Clutch spring removal is made easy with a pair of pointed pliers

8.6 Using a piece of bent metal the clutch can be 'locked'

8.7 Remove the primary chain tensioner

8.8 Take off the engine sprocket and chainwheel in one go

undo the nut on the end of the drive side mainshaft with a suitable socket spanner having beforehand flattened the tab washer behind this nut. With care the rotor can now be levered off its shaft. With motorcycles some years old the shafts and splines are often worn and components such as the alternator rotor come off quite easily. However should any difficulty arise it is advisable to buy a good three leg universal extractor.

4   Now turn to the clutch body. Unscrew the three spring nuts. They may be difficult to slacken because there is a projection on each nut to prevent it working loose.

5   Take out the nuts, springs, spring cups, clutch plates and push rod.

6   Unscrew the clutch centre nut by locking the central body and chain wheel, as shown in the adjacent photograph.

7   Slacken off the chain tensioner screw fully. Remove the engine shaft key (if fitted).

8   The clutch wheel, primary chain and engine sprocket can now be pulled off together if the clutch centre body is first jarred off its taper. If the engine sprocket is tight it may be necessary to use a two/three jaw universe extractor. If the clutch centre body comes away first the 20 clutch roller bearings will be exposed and care must be taken not to let these fall and be lost.

9   Removal of the clutch will expose a plate secured by six screws which gives access to the gearbox sprocket. Remove this plate and unscrew the large nut by first flattening the tab washer and locking the gearbox sprocket. With the nut removed the sprocket will slide off its splines.

10 With the drive components removed attention must now be paid to the timing side of the crankcase before the crankcase can be split.

9   Removal of inner timing side cover. Points assembly. Gearbox internals

1   With the outer timing side cover removed the contact breaker points, kickstart return mechanism, clutch arm and gearshift return spring are exposed.

2   Before removing the two screws holding the contact plate scribe a mark on the plate and housing to assist in reassembly otherwise the ignition will have to be retimed.

3   Take out the central bolt retaining the cam and auto-advance unit. The shaft is threaded to take a 5/16 in UNF extractor bolt or the BSA service tool 61-5005 (early models), 61-3816 (later models), which can be used to extract this mechanism.

4   If the machine is fitted with a revolution counter remove the two bolts holding the cable connection to the front of the timing cover and pull out the revolution counter drive spindle.

5   Pull off the kickstart return spring and anchor plate.

6   Pull off the gear change return spring plate by first slackening the grub screw which locates it. The plate can then be pulled off the gearcharge quadrant spindle.

7   Remove the inner cover by taking out its eleven retaining screws, remembering to tap the cover to break its seal.

8   With the inner timing cover removed the valve timing gear is exposed together with the gearchange selector quadrant and kickstart quadrant. As the gearchange and kickstart quadrants are push fits in the gear box cover these can be removed quite easily.

9   At this stage, if inspection of the gearbox is not required, the crankcases can be split after the timing gears and their associated components are removed. The gearbox internals do not have to be revealed or dismantled for the crankcases to be split. However for the purposes of this manual the gearbox will be dismantled then the timing gears and the crankcase split in that order.

10 Take off the five nuts and spring washers which hold the gear box end cover, break the joint with a hide mallet and remove the cover complete with the gear cluster, cam plate and selector forks.

11 Remove the cam plate plunger and tap out the sleeve pinion from the drive side of the gearbox casting.

9.2 The advance/retard unit behind the contact breaker plate

9.3 Extract the advance unit with a suitable bolt

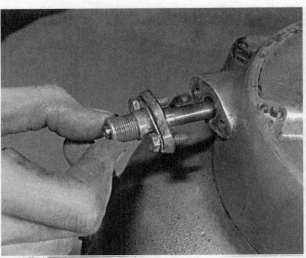

9.4 Remove tachometer drive

9.6 Loosen the grub screw retaining the gearchange return spring plate

9.10 The gear cluster is removed as one unit

9.11 Remember to remove the cam plate plunger

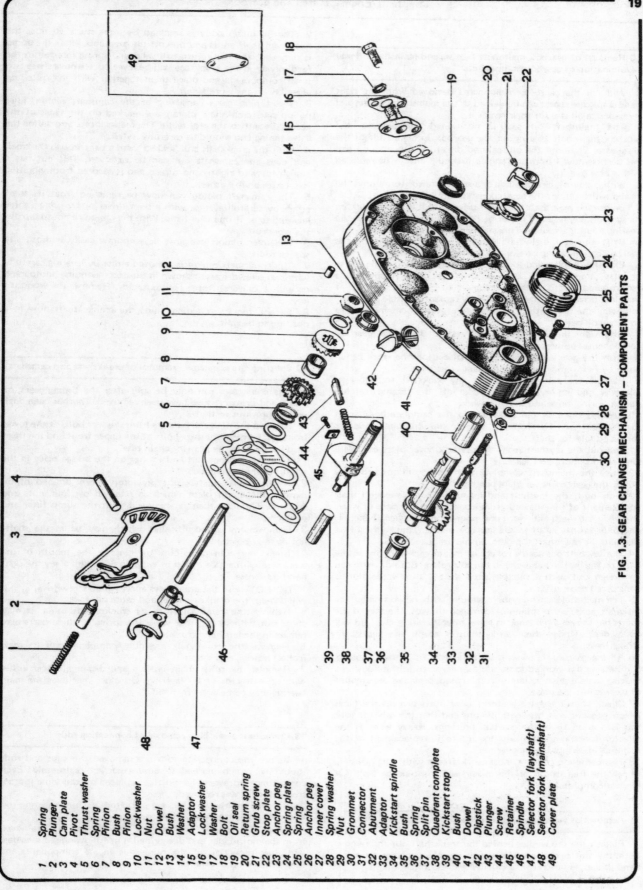

FIG. 1.3. GEAR CHANGE MECHANISM – COMPONENT PARTS

1   Spring
2   Plunger
3   Cam plate
4   Pivot
5   Thrust washer
6   Spring
7   Pinion
8   Bush
9   Pinion
10 Lockwasher
11 Nut
12 Dowel
13 Bush
14 Washer
15 Adaptor
16 Lockwasher
17 Washer
18 Bolt
19 Oil seal
20 Return spring
21 Grub screw
22 Stop plate
23 Anchor peg
24 Spring plate
25 Spring
26 Anchor peg
27 Inner cover
28 Spring washer
29 Nut
30 Grommet
31 Connector
32 Abutment
33 Adaptor
34 Kickstart spindle
35 Bush
36 Spring
37 Split pin
38 Quadrant complete
39 Kickstart stop
40 Bush
41 Dowel
42 Dipstick
43 Plunger
44 Screw
45 Retainer
46 Spindle
47 Selector fork (layshaft)
48 Selector fork (mainshaft)
49 Cover plate

10 Removal of gearbox mainshaft bearing and mainshaft oil seal - examination of gearbox components

1 With all the gearbox components removed from the right hand crankcase access can be gained to the mainshaft bearing and its associate oil seal for their removal.

2 The mainshaft oil seal is positioned adjacent to the gearbox mainshaft bearing on the gearbox sprocket side of the crankcase. Prize out the oil seal and discard it. Once an oil seal has been removed from the engine it should always be replaced with a new one.

3 With a pair of circlip pliers or a small screwdriver remove the circlip which retains the gearbox mainshaft roller bearing.

4 Before the mainshaft bearing is driven out it is preferable that its surrounding housing is warmed with a blow torch so that the bearing housing expands. This makes bearing removal easier.

5 Drift out the mainshaft bearing from the inside of the gearbox casing taking care not to damage the ball race.

6 Examine the bearing for any roughness in the bearing tracks or play, and wash it out in clean paraffin. It should be remembered that it is this bearing which carries all the load of the final drive. If the bearing is faulty replace it with a new part, following the dismantling procedure in reverse, remembering to fit a new oil seal.

7 Attention should now be turned to the other gearbox components for inspection.

8 With the gear cluster intact and placed on the work bench begin by first removing the gear change quadrant.

9 Flatten the tab washer under the kickstart ratchet nut. Unscrew and remove the nut, take off the ratchet, ratchet pinion, spring, bush and thrust washer.

10 Remove the split pin (adjacent to the camplate boss) from the fulcrum pin. Remove the fulcrum pin, complete, selector forks and selector shaft.

11 Pull out the layshaft from the gearbox cover plate complete with all its pinions and gears.

12 Drive the mainshaft complete with all its pinions and gears out of the gearbox cover plate bearing.

13 With both the layshaft and mainshaft complete with their complement of pinions and gears inspect each for signs of wear, chipped gear teeth and worn phosphor-bronze bushes. If there is phosphor-bronze swarf in the gearbox oil then, the layshaft first gear and the mainshaft top gear, bushes are suspect.

14 The layshaft has needle roller bearings at each end, one in the back of the gearbox and one in the coverplate. Both of these can be driven out with a suitable sized drift if their inspection is considered necessary.

15 The mainshaft and layshaft should both be examined for fatigue cracks, worn splines or damaged threads. If either of the shafts has shown a tendency to seize, discolouration involved will be evident. Under these circumstances check the shafts for straightness.

16 All the gearbox bearings must be a tight fit in their housing, if a bearing has worked loose and has revolved in the housing a bearing sealant such as loctite can be used, provided the amount of wear is not too great.

17 Check that the selector forks have not worn on the faces which engages with the gear pinions and that the selector fork rod is a good fit in the gearbox housings. Heavy wear of the selector forks is most likely to occure if replacement of the mainshaft bearings is long overdue.

18 The complete plunger must work freely within its housing. Check the free length of the spring, which should be 2¼" if it has not compressed

11 Removal of valve timing gear, oil pump and worm

1 Flatten the tab washer behind the crankshaft nut and remove both the nut and the worm which is self-extracting on a left hand thread, ie unscrew in a clockwise direction.

2 The oil pump body is retained by three studs. Remove the nuts from their studs and pull off the oil pump. When the pump is removed, the non-return ball valve and spring inserted in the feed oilway in the crankcase will be revealed. Remove these two small components and place them together with the oil pump assembly on the workbench.

3 The timing gears consisting of the camshaft pinion, idler pinion and crankshaft pinion are marked on the faces of the gears adjacent to the gear teeth. The marks locate and define the valve timing and should be carefully noted.

4 With the crankshaft still 'locked' with a bar through the small end eyes the camshaft nut can be removed. This nut has a straightforward right hand thread and is secured from vibrating loose with a tab washer.

5 The camshaft pinion can now be removed from its shaft either by careful leverage with a screwdriver or by using a three jaw extractor if one is at hand. With the pinion removed tap out the woodruff key.

6 The idler pinion can now be withdraw easily without any special tools.

7 The crankshaft pinion is removed either by levering carefully with two screwdrivers as shown in the accompanying photograph or with a universal three jaw extractor. Remove the woodruff key.

8 Except for its retaining studs the crankcase is now in a position to be split.

12 Splitting the crankcase - removal of crankshaft and camshaft

The crankcases can only be split after the timing gears, oil pump and primary drive have been removed. This has been dealt with in previous sections.

1 Remove the nuts and washers from the two bolts at the lower front of the case, the two bolts at the upper front and the three stud nuts within the primary chain case.

There is also one nut and washer on the bridge piece in the mouth of the crankcase.

2 If the crankcase has not previously been drained of oil remove the sump plate which is retained by four nuts and washers. This can then be removed with the sump filter and gasket.

3 Remove any woodruff keys which may be in the shafts noting their positions.

4 Break the crankcase joint by tapping the mouth of the crankcase with a hide mallet or wooden drift. Pull away the right hand crankcase.

DO NOT prise the crankcases apart with a screwdriver as this will damage the crankcase faces and cause oil leaks.

5 Remove the camshaft from its crankcase housing. It may come out with its breather valve and spring or these parts may remain in the left hand crankcase.

6 Remove the crankshaft assembly complete with its main (roller) bearing and spacer.

7 Remove the roller main bearing and distance spacer either with an extractor or by levering. On older machines the main bearing may come away by hand.

13 Crankshaft assembly - removal of connecting rods

1 Before the connecting rods are removed the caps and rods should be punch marked to ensure correct reassembly. Each connecting rod assembly must be replaced on the same journal from which it was removed.

2 Remove the two self locking nuts from each connecting rod assembly and withdraw the big end caps.

These last sections have shown how the engine is removed from the motorcylce and dismantled. The following sections deal with the inspection of the numerous engine components and their renovation or renewal.

10.2 The gearbox main bearing is retained by a circlip

10.2a The bearing is protected by an oil seal

10.10 Remove the split pin from the cam plate spindle

10.13 Check for chipped or broken gear teeth

11.1 The oil pump worm has a left hand thread

11.2 Remove the ball valve which lies beneath the oil pump body

11.3 The valve timing is marked by indentations on the pinions

11.5 Using an extractor to remove the cam wheel

11.7 The mainshaft pinion can be levered off

12.1 The crankcase studs within the primary chaincase

12.4 Using a soft mallet to separate the crankcases

13.1 Mark the connecting rod caps for identification

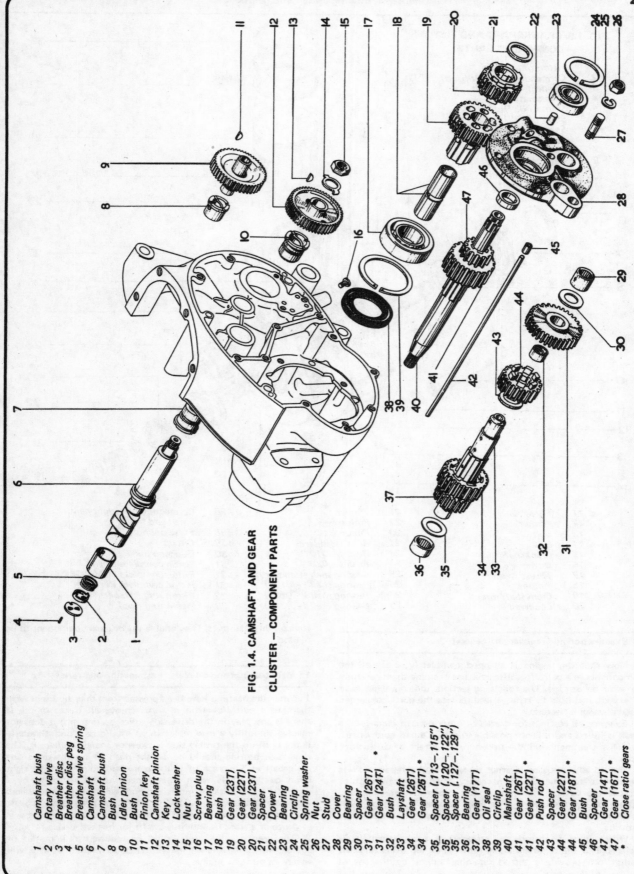

FIG. 1.4. CAMSHAFT AND GEAR CLUSTER — COMPONENT PARTS

1   Camshaft bush
2   Rotary valve
3   Breather disc
4   Breather disc peg
5   Breather valve spring
6   Camshaft
7   Camshaft bush
8   Bush
9   Idler pinion
10 Bush
11 Pinion key
12 Camshaft pinion
13 Key
14 Lockwasher
15 Nut
16 Screw plug
17 Bearing
18 Bush
19 Gear (23T)
20 Gear (22T)
20 Gear (23T) *
21 Spacer
22 Dowel
23 Bearing
24 Circlip
25 Spring washer
26 Nut
27 Stud
28 Cover
29 Bearing
30 Spacer
31 Gear (26T)
31 Gear (24T) *
32 Bush
33 Layshaft
34 Gear (26T)
34 Gear (26T) *
35 Spacer (.113—.115")
35 Spacer (.120—.122")
35 Spacer (.127—.129")
36 Bearing
37 Gear (17T)
38 Oil seal
39 Circlip
40 Mainshaft
41 Gear (26T)
41 Gear (22T) *
42 Push rod
43 Spacer
44 Gear (22T)
44 Gear (18T) *
45 Bush
46 Spacer
47 Gear (14T)
47 Gear (16T) *
*   Close ratio gears

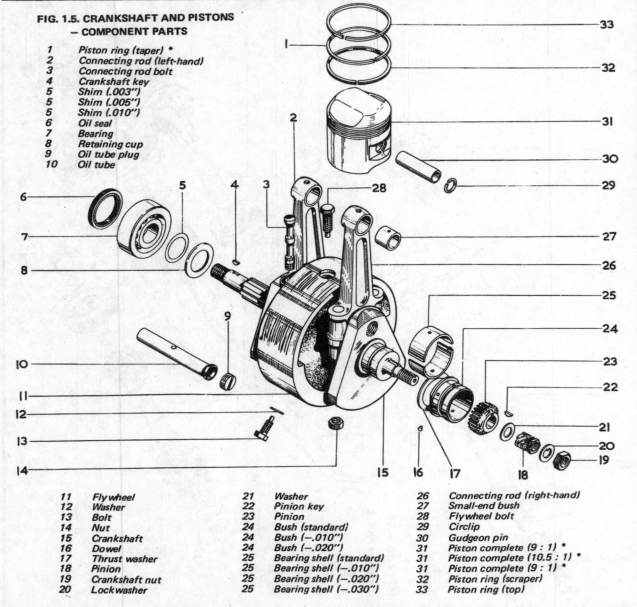

**FIG. 1.5. CRANKSHAFT AND PISTONS
– COMPONENT PARTS**

1	Piston ring (taper) *
2	Connecting rod (left-hand)
3	Connecting rod bolt
4	Crankshaft key
5	Shim (.003")
5	Shim (.005")
5	Shim (.010")
6	Oil seal
7	Bearing
8	Retaining cup
9	Oil tube plug
10	Oil tube

11	Flywheel	21	Washer	26	Connecting rod (right-hand)
12	Washer	22	Pinion key	27	Small-end bush
13	Bolt	23	Pinion	28	Flywheel bolt
14	Nut	24	Bush (standard)	29	Circlip
15	Crankshaft	24	Bush (—.010")	30	Gudgeon pin
16	Dowel	24	Bush (—.020")	31	Piston complete (9 : 1) *
17	Thrust washer	25	Bearing shell (standard)	31	Piston complete (10.5 : 1) *
18	Pinion	25	Bearing shell (—.010")	31	Piston complete (9 : 1) *
19	Crankshaft nut	25	Bearing shell (—.020")	32	Piston ring (scraper)
20	Lockwasher	25	Bearing shell (—.030")	33	Piston ring (top)

14 Examination and renovation - general

1 Now that the engine is stripped completely, clean all the components in a petrol/paraffin mix and examine them carefully for wear or damage. The following sections indicate what wear to expect and how to remove and replace the parts concerned when renewal is necessary.

2 Examine all castings for cracks or other signs of damage. If a crack is found and it is not possible to obtain a new component, specialist treatment will be necessary to effect a satisfactory repair.

3 Should any studs or internal threads require repair, now is the appropriate time. Care is necessary when withdrawing studs because the casting may not be too strong at certain points. Beware of overtightening; it is easy to shear a stud by over-tightening giving rise to further problems, especially if the stud bottoms.

4 Where internal threads are stripped or badly worn, its is preferable to use a thread insert rather than tap oversize. Most dealers can provide a thread retaining service by the use of Helicoil thread inserts. They enable the original component to be re-used.

15 Main bearings and oil seals - examination and renovation

1 When the bearings have been pressed from their housings wash them in a petrol/paraffin mix to remove all traces of oil. If there is any play in the drive side roller bearing or if it does not revolve smoothly a new replacement should be fitted. Similarly if the bush on the timing side is worn or scored, replace it. The drive side bearing should be a tight push fit on the drive side mainshaft and a press fit in the crankcase housing. If the right hand crankshaft is worn it will need to be reground by a specialist and a suitable undersized bush will have to be supplied. These are available in —.010" to 0.020", A proprietary sealant such as Loctite can be used to secure the bearings if there is evidence of a slack fit and they are fit for further service.

2 The crankcase oil seal should be renewed as a matter of course, whenever the engine is stripped completely. This will ensure an oil tight engine.

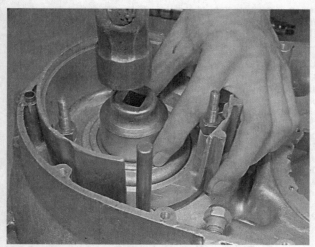
15.2 Pressing in a new oil seal

orifice. Note that the retainer plug is retained by a centre punch mark and that it will be necessary to use an impact screwdriver after the indentation has been drilled out.

6 Wash the oil tube in a petrol/paraffin mix and check that all the internal drillings in the crankshaft are quite free before the tube is replaced. A jet of compressed air is best for this purpose. Make sure the retaining plug is tightened fully and centre punch the crankshaft at the screw slot, to retain the plug in position.

17 Camshaft and timing pinion bushes - examination and renovation

1 It is unlikely that the camshaft and timing pinion bushes will require attention unless the machine has covered a high mileage. The normal wear rate is very low. Bushes in the right hand crankcase can be removed by heating the crankcase to expand the surrounding metal and driving them out from the outside, using a two-diameter dirft of the correct size. Fit the new bushes whilst the crankcase is still hot and make sure they are correctly aligned so that any oil feed holes register with those of the crankcase.

2 The blind bushes in left hand crankcase are more difficult to remove. The recommended technique is to tap the bushes with a Whitworth thread and screw home a bolt of matching thread. If the crankcase is now heated, the bolt head can be gripped in a vice and the crankcase driven off the bush with a rawhide mallet. When the camshaft bush is replaced care must be taken to engage the peg with the breather porting disc that lies behind the bush.

3 The camshaft bush is machined from scintered bronze and only the smallest amount of metal will need to be removed after they are pressed into position. The correct internal box diameters (fitted) are as follows:

Left hand crankcase   0.8115 - 0.8125 inch
Right hand crankcase   0.8750 - 0.8760 inch

The intermediate (idler) timing gear bush is machined from phosphor-bronze and should have an internal bore diameter fitted of 0.6885" - 0.6875".

16 Crankshaft assembly - examination and renovation

1 Wash the complete crankshaft assembly (if it has not already been dismantled) with a petrol/paraffin mix to remove all surplus oil. Mark each connecting rod and cap to ensure they are replaced in exactly the same position, then remove the cap retainer nuts so that the caps and connecting rods can be withdrawn from the crankshaft. Keep the bolt and nuts together in pairs so tha they are replaced in their original order. It is best to unscrew the nuts a turn at a time to obviate the risk of distortion.

2 Inspect the bearing surfaces for wear. Wear usually takes the form of scuffing or scoring, which will immediately be evident. Bearing shells are cheap to renew; it is wise to renew the shells if there is the slightest question of doubt about the originals.

3 More extensive wear will require specialist attention either by having the crankshaft reground or by fitting a service exchange replacement.

   If the crankshaft is reground, two undersizes of bearing shells can be obtained: -0.010 in and -0.020 in. The following table gives details of the various sizes of shell bearing and crankshaft, in relation to one another.

Shell bearing	Crankshaft size (limits)	
	in.	mm
Standard	1.6865-1.6870"	42.837-42.849 mm
Undersize		
—0.010	1.6765-1.6770"	42.583-42.595 mm
—0.020	1.6665-1.6670"	42.329-42.341 mm

   It is particularly important to note that the white metal bearing shells are prefinished to give the correct diametrical clearance and on no account should the bearings be scraped or the connecting rod and cap joint be filed in order to achieve a satisfactory fit. If such action seems necessary the crankshaft has not been reground to the correct tolerances.

4 Check all the connecting rod bolts for stretch by comparing them with a new bolt. When the bolts are tightened fully to the recommended setting stretch should not exceed 0.005 in.

5 It is not usually necessary to disturb the crankshaft assembly unless the lubrication system has become contaminated in which case it may be advisable to clean out the central oil tube. Access is gained by unscrewing the retainer plug found in the right hand end of the crankshaft and then removing the flywheel bolt adjacent to the big end journal. The oil tube can be hooked out by passing a length of rod through the flywheel bolt

18 Camshaft, tappet followers and timing pinions - examination and renovation

1 Examine the camshaft, checking for wear on the cam form which is usually evident on the opening flank and on the lobe. If the cams are grooved or if there are scuff or score marks that cannot be removed by light dressing with an oilstone, the camshaft should be renewed.

2 When extensive wear has necessitated the renewal of the camshaft, the camshaft and tappet followers should be renewed at the same time. It is a false economy to use the existing camshaft followers with a new camshaft since they will promote a more rapid rate of wear.

3 Check the camshaft pinion and idler pinion for worn or broken teeth. Damage is most likely to occur if some engine component has failed during service and particles of metal have circulated with the lubrication system. Excessive backlash in the pinions will lead to noisy timing gear.

19 Cylinder barrel - examination and renovation

1 There will probably be a lip at the uppermost end of each cylinder bore that denotes the limit of travel of the top piston ring. The depth of the lip will give some indication of the amount of bore wear that has taken place, even though the amount of wear is not evenly distributed.

2 Remove the rings from the pistons, taking great care as they are brittle and easily broken. Most wear occurs within the top half of the bore, so the pistons should be inserted and the clearance between the skirt and cylinder wall measured. If measurement by feeler gauge shows the clearance is 0.005" greater or more than the figure quoted in the specifications section, the

cylinder is due for a rebore. Oversize pistons are supplied in three sizes: +0.010'' +0.020'' and 0.040''; the cylinders should be rebored to suit, as shown in the accompanying list of tolerances:

650 cc models

Piston size		Bore size	
in.	mm	in.	mm
Standard		2.9521	74.9830
		2.9530	75.0062
+0.010	+0.254	2.9621	75.237
		2.9630	75.2602
+0.020	+0.508	2.9721	75.491
		2.9730	75.5142
+0.040	+1.016	2.9921	75.999
		2.9930	76.022

500 cc models

Piston size		Bore size	
in.	mm	in.	mm
Standard		2.5780	65.481
		2.5790	65.506
+0.010	+0.254	2.5880	65.735
		2.5890	65.760
+0.020	+0.508	2.5980	65.989
		2.5990	66.014
+0.040	+1.016	2.6180	66.497
		2.6190	66.522

3 Give the cylinder barrel a close visual inspection. If the surface of either of the bores is scored or grooved, indicative of a previous engine seizure or a displaced circlip and gudgeon pin, or rebore is essential. Compression loss will have a very marked effect on performance.

4 Check that the outside of the cylinder barrel is clean and free from road dirt. Use a wire brush on the cooling fins if they are obstructed in any way. The application of matt cylinder black will help improve the head radiation.

5 Check that the base flange is not cracked or damaged. If the engine has been overstreesed, one of the first parts to fail is the base of the cylinder barrel, either at the holding down points or around the base of each bore. If a crack is found, the cylinder barrel must be renewed.

6 The rebore limit is 0.040''. Above this size, the cylinder walls cannot be considered to have sufficient thickness consistent with safety and reliability. Re-sleeving or a service exhange replacement is the only practicable solution to the problem.

20 Pistons and piston rings - examination and renovation

1 Attention to the pistons and piston rings can be overlooked if a rebore is necessary, since new replacement will be fitted.

2 If a rebore is not considered necessary, examine each piston closely. Reject pistons that are scored or badly discoloured as the result of exhaust gases by-passing the rings.

3 Remove all carbon from the piston crowns, using a blunt scraper which will not damage the surface of the piston. Clean away all carbon deposits from the valve cutaways and finish off with metal polish so that a clean, shining surface is achieved. Carbon will not adhere so readily to a polished surface.

4 Check that the gudgeon pin bosses are not worn or the circlip grooves damaged. Check that the piston ring grooves are not enlarged. Side pleat should not exceed 0.003''.

5 Piston ring wear can be measured by inserting the rings in the bore from the top and pushing them down with the base of the piston so thay they are square in the bore and about 1½'' down. If the end gap exceeds 0.014'' (all rings) renewal is necessary.

6 Check that there is no build up of carbon on the inside surface of the rings or in the grooves of the pistons. Any build up should be removed by careful scraping.

7 The piston crowns will show whether the engine has been reboxed on some previous occasion. All oversize pistons have the

rebore size stamped on the crown. This information is essential when ordering replacement piston rings.

21 Small end bearings - examination and renovation

1 The amount of wear in the small end bushes can be ascertained by the fit of the gudgeon pins. The pin should be a good sliding fit in each case, without evidence of any play. Renewal can be effected by using a simple drawbolt arrangement (as illustrated) whereby the new bush is used to press the old bush out of location.

2 It is essential to ensure the oilway in the bush locates with the oilway in the connecting rod, otherwise the bearing will run dry and rapid wear will occur.

3 After the bushes have been fitted they will have to be reamed ut to the correct size. Cover the mouth of the crankcase with rag to prevent metallic particles from dropping in and ream out to 0.7503'' - 0.7506''.

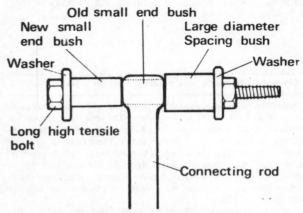

Fig. 1.6. Removal of small end bush

22 Cylinder head and valves - dismantling, examination and renovation

1 It is best to remove all carbon deposits from the combustion chambers, before removing the valves for grinding-in. Use a blunt-ended scraper so that the surface of the combustion chambers is not damaged and finish off with metal polish to achieve a smooth, shiny surface.

2 Before the valves can be removed, it is necessary to obtain a valve spring compressor of the correct size. This is necessary to compress each set of valve springs in turn, so that the split collets can be removed from the valve cap and the valve and valve spring assembly released. Keep each set of parts separate; there is no fear of inadvertently interchanging the valves because the heads are marked 'IN' or 'EX'.

3 Before giving the valves and valve seats further attention, check the clearance between each valve stem and the valve guide in which it operates. Some play is essential in view of the high temperatures involved, but if the play appears excessive, the valve guides must be renewed.

4 To remove the old valve guides, heat the cylinder head and drive them out of position with a double diameter drift of the correct size. Replace the new guides, whilst the cylinder head is still warm.

5 Grinding in will be necessary, irrespective of whether new valve guides have been fitted. This action is necessary to remove the indentations in the valve seats caused under normal running conditions by the temperatures within the combustion chambers. It is also necessary when new valve guides have been

fitted, in order to re-align the face of each valve with its seating.

6 Valve grinding is a simple task. Commence by smearing a trace of fine valve grinding compound (carborundum paste) on the valve seat and apply a suction tool to the head of the valve. Oil the valve stem and insert the valve in the guide so that the two surfaces to be ground in make contact with one another. With a semi-rotary motion, grind in the valve head to the seat, using a backward and forward action. Lift the valve occasionally so that the grinding compound is distributed evenly. Repeat the operation until a ring of light grey matt finish is obtained on both valve and seat. This denotes the grinding operation is complete. Before passing to the next valve, make sure that all traces of compound have been removed from both the valve and its seat and that none has entered the valve guide. If this precaution is not observed, rapid wear will take place due to the abrasive nature of the carborundum base.

7 When deeper pit marks are encountered, or if the fitting of new valve guide makes it difficult to obtain a satisfactory seating, it will be necessary to use a valve seat cutter set to an angle of 45° and a valve refacing machine. This course of action should be resorted to, only in an extreme case, because there is risk of pocketing the valve and reducing performance. If the valve itself is badly pitted, fit a replacement.

8 Before reassembling the cylinder head, make sure that the split collets and the taper with which they locate on each valve are in good condition. If the collets work loose whilst the engine is running, a valve will drop and cause extensive engine damage. Check the free length of the valve springs with the specifications section and renew any that have taken a permanent set.

9 Reassemble by reversing the procedure used for dismantling the valve gear. Do not neglect to oil each valve stem before the valve is replaced in the guide.

10 Before setting aside the cylinder head for reassembly, make sure that the cooling fins are clean and free from road dirt. Check that no cracks are evident, especially in the vicinity of the holes through which the holding down studs and bolts pass, and near the spark plug threads.

11 Finally make sure that the cylinder head flange is completely free from distortion at the joint it makes with the cylinder barrel. An aluminium alloy cylinder head will distort with comparative ease if it is tightened unevenly and may lead to a spate of blowing cylinder head gaskets. If the amount of distortion is not too great flatness can be restored by carefully rubbing down on a sheet of fine emery cloth wrapped around a sheet of plate glass. Otherwise it may be necessary to have the cylinder head flange refaced by a machining operation.

22.2 Removing the valve springs with a valve spring compressor

23 Tappet follower - examination and renovation

1 Mention has not been made of the tappets or tappet guide block which seldom require attention. The amount of wear within the tappet block can be ascertained by rocking the tappet whilst it is within the tappet block. It should be a good sliding fit, with very little sideways movement.

2 To remove and replace the tappet block, first remove the locking screw. The block can be drifted out of position.

24 Pushrods, rocker spindles and rocker arms - examination and renovation

1 Check the pushrods for straightness by rolling them on a sheet of glass. If they are bent they should be renewed since it is not easy to effect a satisfactory repair.

2 Check the end pieces to ensure that they are a tight fit on the light alloy tubes. If the end pieces work loose the pushrod must be renewed. It is unlikely that the end pieces will show signs of wear at the point where they make contact with the rocker arms and the tappet followers, unless the machine has covered a very high mileage. Wear usually takes the form of chipping or breaking through the hardening, which will necessitate renewal.

3 Examine the tips of the rocker arms. If wear is evident, both the valve clearances adjusters and the ball pins should be renewed. The latter should be pressed into place with the drilled flat towards the rocker spindle.

4 If it is necessary to renew the rocker spindles they can be driven out of the rocker housing by means of a soft metal drift after the rocker spindle nuts have been unscrewed and removed. Fig. 1.1 shows the way in which the various washers are assembled. Before fitting the spindles check that the oilways are clean and unobstructed preferably by using a jet of compressed air. Lubricate the spindle thoroughly before it is inserted. There is no adjustment for endfloat. This function is performed by the spring washers fitted between each end of the rocker arm and the rocker box.

25 Engine reassembly - general

1 Before the engine/gearbox is reassembled, all the various components must be cleaned thoroughly so that all traces of old oil, sludge, dirt and gaskets etc., are removed. Wipe each part with clean, dry lint free rag so that there is nothing to block the internal oilways of the engine during reassembly.

2 Make sure that all traces of the old gaskets have been removed and that the mating surfaces are clean and undamaged. One of the best ways to remove old gasket cement is to apply a rag soaked in methylated spirit. This acts as a solvent and will ensure the cement is removed without resort to scraping and subsequent risk of damage.

3 Gather together all the necessary tools and have available an oil can filled with clean engine oil. Make sure the new gaskets and oil seals are to hand; nothing is more infuriating than having to step in the middle of a reassembly sequence because a vital gasket or replacement has been overlooked.

4 Make sure that the reassembly area is clean and that there is adequate working space. Refer to the torque and clearance setting wherever they are specified. Many of the smaller bolts are easily sheared if they are overtightened. Always use the correct size screwdriver for the crosshead screws and never an ordinary screwdriver or punch.

26 Engine reassembly - rebuilding the crankshaft

1 Refit the connecting rods to the crankshaft assembly in their original positions, using the marks made during the dismantling operation as a guide. Make sure that the shell bearings are

located correctly, then replace the end caps, the retaining nuts and bolts. Tighten each nut evenly until the connecting rods and their caps seat correctly, then tighten the nuts with a torque wrench to a load of 22 ft. lb. Check that both connecting rods revolve freely, with absolute freedom from play.

2 Apply a pressure oil can to the drilling at the right hand end of the crankshaft and pump until oil is expelled from both big ends. This is essential, to ensure that the oil passages are free from obstruction and full of oil.

3 If a torque wrench is not available, tighten the connecting rod nuts until a micrometer reading shows the bolts have stretched to a maximum of 0.005". This extension figure is to be preferred as a means of accurately tensioning the bolts without risk of overstress.

27 Engine reassembly - reassembling the crankcases

1 If the main bearings have been removed, replace the large bush type bearing in the right hand crankcase and the outer race of the roller bearing in the left hand crankcase. It is advisable to heat the respective crankcases beforehand so that the bearings will drop into place without difficulty.

2 The inner race of the roller bearing should be driven on to the left hand side of the crankshaft assembly until it is hard against the shoulder of the crankshaft.

3 Any shims that have been fitted to limit the end float of the crankshaft assembly must be fitted behind the main roller bearing adjacent to the crankshaft.

4 Mount the left hand crankcase on two blocks of wood so that there will be sufficient clearance for the end of the crankshaft to project downwards without touching the bench. Lubricate the main bearings and the camshaft bushes and place the rotary breather valve and spring in the camshaft bush. If the camshaft is not fitted to the right hand crankcase, position it in its left hand bush taking care that the slot in the end of the camshaft engages with the projection of the rotary breather valve.

5 Lower the crankshaft assembly into position and give it a sharp tap to ensure that the inner race of the roller bearing is fully engaged with the outer race.

6 Coat the jointing face of the right hand crankcase with gasket cement and lower into position, after checking that the connecting rods are centrally disposed. If the camshaft is still attached to the right hand crankcase it will be necessary to rotate it until it engages with the rotary breather valve before the crankcases will meet. Push both crankcases together so that they mate correctly all round and check that the crankshaft and camshaft revolve quite freely before the securing bolts and studs are replaced and tightened by hand. Do not omit the thrust washer between the crankshaft and timing side main bearing.

7 Check that the cylinder barrel junction of the crankcase is level and if necessary adjust by light tapping. When a level surface is achieved and the crankshaft and camshaft are free to revolve, the securing bolts can be tightened fully with a torque wrench to 20 ft. lb (studs) and 13 ft. lb. bolts.

28 Engine reassembly - replacing the timing pinions, oil pump, sump plate and filter

1 Replace the sump plate and filter using a new gasket cemented on both sides and replace and tighten its four retaining washers and nuts.

2 Replace the woodruff key in the 'right hand' crankshaft and refit the crankshaft pinion with the valve timing mark facing on the outside.

3 Replace the crankshaft distance piece adjacent to the crankshaft pinion and then screw on the oil pump worm gear which has a left-hand thread and is therefore replaced by screwing anti-clockwise. Place the tab washer in position and secure with the nut which also has a left hand thread: then tighten with a torque wrench to 60 lb. ft.

4 Place a new oil pump gasket in position and replace the

26.1 Use a torque wrench when finally tightening the connecting rod bolts

27.4 Do not omit the rotary valve breather components

27.6 Do not omit the timing side thrust washer

28.4 Remember to replace the ball valve

28.5 The cam wheel is located by a Woodruff key

28.5a Do not omit the tab washer underneath the camshaft nut

spring and ball valve in its recess which is at the lower right hand side of the uppermost oil pump body mounting stud. Carefully slide the oil pump body over the three mounting studs taking care when the pump worm engages with the worm gear. Replace the nuts and washers of each stud and tighten down evenly, finally to a torque setting of 7 ft. lb.

5   Replace the woodruff key of the camshaft and tap on the camshaft pinion making sure that the valve timing mark is on the outside. Replace the tab washer and camshaft nut. Tighten the camshaft not fully and turn the tab washer over the nut after tightening.

6   The idler pinion can now be replaced and the valve timing marks lined up.

7   If the gear cluster has not yet been replaced it can now be assembled.

Fig. 1.7. Aligning the valve timing marks

29 Gearbox reassembly - replacing gearbox components and inner timing covers

1   It will be assumed that all bearings, bushes or oil seals have been renewed as necessary as recommended from Section 10.13 onwards of this Chapter.

2   Press the mainshaft through its bearing in the outer cover making sure that the distance piece is in position between the bearing and the small (1st) gear. Grip the shaft in a vice using soft metal or wooden clamps and replace the kickstart ratchet mechanism consisting of, the ratchet gear washer, bronze bush, spring, ratchet pinion, ratchet, tab washer and nut, in that order. Turn the washer over the nut after tightening with a torque wrench to 60 ft.lbs.

3   Place one of the thrust washers in position, adjacent to the layshaft needle roller bearing, on the inside face of the cover. Position the layshaft first gear with the dogs facing inwards and insert the layshaft complete with its sliding gear second and top gear.

4   Position the layshaft selector fork, slide the mainshaft sliding gears onto the shaft and position the mainshaft selector fork.

5   Insert the camplate through the slot in the cover with the long end of the outer track at the bottom. Insert the camplate fulcrum pin and lock it in position with its split pin. Insert the selector fork spindle through the selector forks and into its recess in the outer cover.

6   If the mainshaft top gear has not already been fitted press the gear through its bearing and replace the gearbox sprocket securing it with the locknut and tab washer.

7   Locate the camplate plunger and spring in their recess at the back of the gearbox.

8   Pick up the gearbox outer cover complete with the gear cluster and place the other thrust washer over the end of the layshaft and carefully slide the whole assembly into the box.

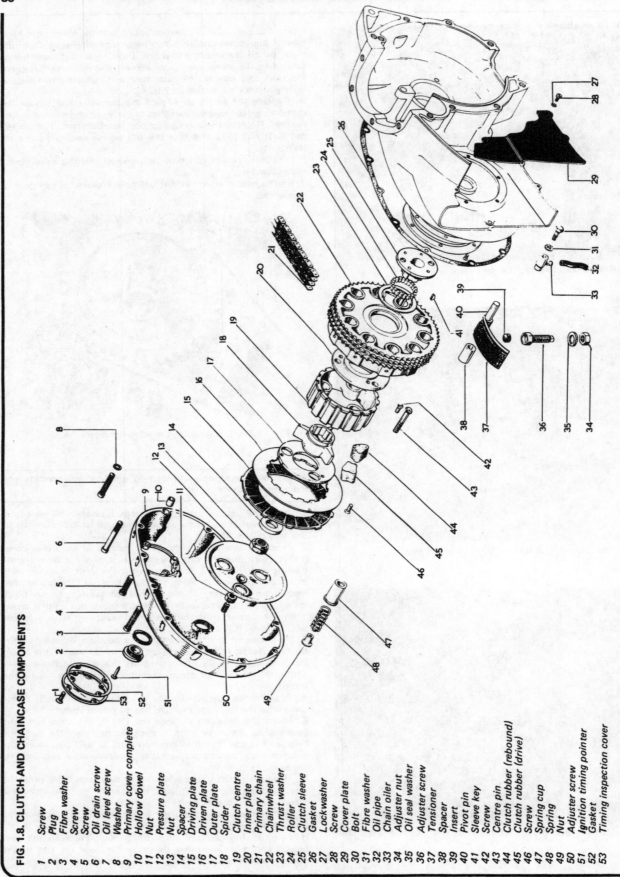

FIG. 1.8. CLUTCH AND CHAINCASE COMPONENTS

1 Screw
2 Plug
3 Fibre washer
4 Screw
5 Screw
6 Oil drain screw
7 Oil level screw
8 Washer
9 Primary cover complete
10 Hollow dowel
11 Nut
12 Pressure plate
13 Nut
14 Spacer
15 Driving plate
16 Driven plate
17 Outer plate
18 Spider
19 Clutch centre
20 Inner plate
21 Primary chain
22 Chainwheel
23 Thrust washer
24 Roller
25 Clutch sleeve
26 Gasket
27 Lockwasher
28 Screw
29 Cover plate
30 Bolt
31 Fibre washer
32 Oil pipe
33 Chain oiler
34 Adjuster nut
35 Oil seal washer
36 Adjuster screw
37 Tensioner
38 Spacer
39 Insert
40 Pivot pin
41 Sleeve key
42 Screw
43 Centre pin
44 Clutch rubber (rebound)
45 Clutch rubber (drive)
46 Screw
47 Spring cup
48 Spring
49 Nut
50 Adjuster screw
51 Ignition timing pointer
52 Gasket
53 Timing inspection cover

9 To facilitate the meshing of the top gear pinions revolve the shafts gently. Do not attempt to force the cover home. With the cover in position secure it with the five nuts and washers.

10 Position the camplate midway in the slot and insert the small end of the gear selector quadrant into the small hole in the cover at the same time engaging the quadrant plunger in the camplate.

11 The gearbox is now in a position for the inner timing cover to be replaced. Ensuring that all the cover faces are clean and grease free smear jointing compound over the outer (right hand) crankcase faces and also on the inside of the inner timing cover. Place a new gasket in position taking care to line up the holes of the paper gasket with the holes in the crankcase (for accepting the timing cover retaining screws). Position the inner timing cover over the right hand crankcase and press into place making sure that the kickstart quadrant engages and that the gear change quadrant spindle slides through the 'sleeve' in the inner timing cover by replacing and tightening the eight crosshead screws round the outer edge, the two in the centre and the one under the clutch lever.

12 If the machine is fitted with a revolution counter replace the revolution counter drive spindle, its connector and the two bolts holding the cable connection to the front of the timing cover.

13 Replace the footchange return spring and stop plate on the gear change quadrant spindle and lock the stop plate in position by tightening the adjacent grub or set screw.

---

30 Reassembly of drive side components - clutch, engine sprocket alternator - primary chain and chaincase cover

---

1 If the gearbox sprocket has not already been replaced now is the stage at which to do so. Slide the sprocket over its splines and secure it in position with its large tab washer and nut. Tighten the nut fully and flatten the tab washer over the nut.

2 Replace the circular plate at the back of the chaincase. Use a new gasket and smear jointing compound on the mating faces then secure the plate with its six screws.

3 Replace the woodruff key in the clutch mainshaft and tap on the clutch sleeve.

4 With the engine lying on its right side grease the bearing surface of the clutch sleeve and position the clutch bearing thrust washer concentric with the box of the clutch sleeve.

5 Now take the clutch chain wheel, the engine sprocket primary chain and engine sprocket distance piece. Slide the distance piece over the drive side mainshaft. Place the primary chain around the engine sprocket and clutch chainwheel and picking up the engine sprocket, chainwheel and chain, slide both sprockets and chain over their respective shafts at the same time. Carefully replace the twenty rollers between the clutch sleeve and the bearing surface of the clutch chainwheel and slide the clutch central body over the splines of the clutch sleeve. Secure the clutch by replacing the distance piece, recessed side outwards, then secure the self locking nut tightened to a torque of 70-75 lb/ft.

6 On models fitted with energy transfer ignition equipment the timing disc is fitted next to the engine sprocket with its peg facing outwards at approximately 9 o'clock and the piston at top dead centre. The rotor for these models has two holes at the rear, the one marked 'S' is used for both A50 and A65 models. Locate the peg in the appropriate hole and secure the rotor with its nut and washer turning the washer onto the nut after tightening.

On models with battery ignition equipment replace the woodruff key in the crankshaft, replace the rotor and secure with a nut and washer tightened to a torque of 60 ft lb.

7 Replace the clutch plates sliding one segmented driving plate into the clutch chainwheel first and then one plain plate and so on until all the plates are assembled, there being six of each.

8 Insert the clutch pushrod into the hollow gearbox mainshaft making sure that is 'meshes' with the clutch arm, plunger and ball on the right hand side of the crankcase. Replace the pressure plate, spring cups, springs and nuts. The nuts can be screwed in with a screwdriver some of the way but a sharp nosed pair of

pliers will be needed to screw them down until their faces are flush with the end of the clutch studs, which is their normal setting. The springs must be tightened evenly so that the clutch plates are parallel when the clutch is lifted.

9 Replace the primary chain tensioner and tighten the chain using the adjuster screw until there is about 1/8'' up and down movement on the top run of the chain. Then lock the chain tensioner by tightening the cap nut.

10 Place the stator over the three studs so that the leads are on the outside at approximately 2 o'clock and replace the three self lockings nuts. Tighten to a torque of 10-15 ft. lb.

It is import nt that there is a gap between the rotor and stator of 0.015'' or more. If there is any variation due to the studs having been displaced, the studs should be very carefully set over. The gap can be checked with feeler gauges between the stator pole pieces and the rotor.

11 Pass the lead from the alternator through the cable guide and crankcase wall to pass underneath the engine.

12 Apply jointing compound to both faces of the chaincase and, using a new gasket, replace the cover tightening the screws evenly to avoid distortion. See that the oil level and drain screws are correctly positoned on the lower run with aluminium washers under the heads of the screws.

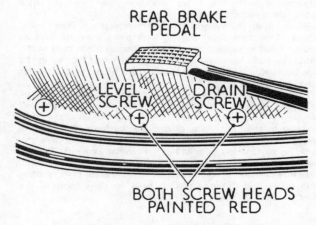

Fig. 1.9. Primary chaincase oil level and drain screws

30.5 Using a piece of bent metal to lock the clutch when tightening the clutch centre nut

31 Engine reassembly - replacing pistons and cylinder barrel

1 Before fitting the pistons, it is advisable to pad out the crankcase mouth with clean rag, in order to prevent a displaced circlip from falling in. Extra, unnecessary dismantling work may be called for if the worst happens and this precaution is not observed.
2 Oil both small ends, the bosses of each piston and the gudgeon pins. Fitting is made easier if the pistons are warmed first, especially when new pistons are being fitted after a rebore.
3 As each gudgeon is fitted, check that the circlips have engaged with their retaining grooves. A misplaced circlip can work free and cause extensive engine damage, especially if the gudgeon pin is allowed to work out of position and score the cylinder bore.
4 Always use new circlips. It is false economy to use the old components, even if they appear perfect. There is too much to risk if a circlip breaks or works free whilst the engine is running.
5 Standard pistons can be replaced either way round because the valve cutaways in the crown are equal. However, they should always be replaced in their original position if possible, hence the need to mark inside the skirt when dismantling. High compression pistons must be fitted so that the greater cutaway corresponds with the larger diameter inlet valves, unless specifically stated otherwise.
6 When fitting the cylinder block it is necessary to position the pistons at bottom dead centre, with some form of support beneath them to prevent them from tilting. A slight taper at the bottom of each bore will act as a lead-in when feeding in the rings by hand; it is preferable to use a pair of piston ring clamps if they are available.
7 Fitting the cylinder block is also a two man task, in view of the weight of this component. Smear each bore with engine oil, then gently lower the block until the piston rings enter the bores correctly or the piston ring clamps are displaced. Remove the clamps and any rag from the crankcase mouth, then lower the block on to the new base gasket (no gasket cement). Tighten down with the cylinder base nuts and their shakeproof washers.
8 Before proceeding further, check that the crankshaft revolves quite freely and that the pistons have no tight spots in the bores.

32 Engine reassembly - replacing pushrods and cylinder head

1 Place a new cylinder head gasket in position and slacken off the inlet valve rockers completely. Place the head in position over the four studs, screw in the five cylinder head bolts and replace the four nuts and washers on the studs. Tighten down each a little at a time criss-crossing the head from the centre outwards to minimize distortion. Do not forget the short bolt inside the push rod tower; this bolt cannot be fitted with the push rods in situ. Using a torque wrench tighten the bolts to 25 ft. lb and the nuts to 26 ft. lb.
2 When the cylinder head is finally torqued down replace the two short push-rods on the two outer tappets and under the inlet (rear) rocker arms. Assemble the exhaust (front) rockers in the order detailed in Fig. 1.1 with the adjuster screws over the valves and fit the two long pushrods on the two inner tappets and under the exhaust rocker ball pins. With the exhaust rocker assembly in place replace and tighten the rocker spindle nut.

33 Engine reassembly - tappet adjustment

1 To set the tappets (or valve clearance) the valve must be in the correct position, that is with the cam follower (tappet) on the base circle of the cam as follows:
    Left-hand inlet valve spring fully compressed. Set the right-hand inlet valve.
    Right-hand inlet valve spring fully compressed. Set the left-hand inlet valve.
    Left-hand exhaust valve spring fully compressed. Set the right hand exhaust valve.
    Right-hand exhaust valve spring fully compressed. Set the left-hand exhaust valve.
2 With the valve in the correct position check the gap with the appropriate feeler gauge. 0.008 in inlet clearance and 0.010 in. exhaust clearance.
    If adjustment is required slacken off the locknut and screw the square adjuster in or out as necessary. Always recheck the setting after tightening the locknut.
3 Clean the mating surfaces of the rocker box and cover and apply some jointing compound. Position a new gasket and replace the rocker box cover tightening it down with its six domed nuts and washers.

34 Replacing the contact breaker assembly and timing the ignition

1 Replace the automatic advance/retard mechanism by inserting it in the extension of the idler gearwheel. Secure the mechanism and contact breaker cam by securing the centre bolt into the idler gear. Replace the contact breaker plate in its housing and secure it by inserting and tightening its two retaining screws.
2 It is convenient, whilst the engine is still on the bench to check and if necessary test the ignition timing. Different procedures are required according to whether or not a stroboscope is available. Refer to Chapter 3, Section 7 for the relevant detailed information.
3 The static method of ignition is preferred amongst owners even though it is less occurate. Comparatively few have access to a stroboscope or instruction in its use. Provision is made in the engine design to make timing by the static method as simple yet as accurate as possible.

35 Replacing the outer timing cover, gear and kickstart levers

1 If the clutch cable connector, abutment and adaptor were removed to release the clutch cable when the engine was originally stripped down replace them in their housing at the back of the inner timing cover, remembering to connect the spring loaded nipple of the connector with the clutch lever cam.
2 Reconnect the wire to the contact breaker unit. There are colour coded for identification purposes. The upper set of points serves the right-hand cylinder and is connected by the black/yellow wire (early models, 1969 on wiring may differ); the remaining wire (black/white, early models) is connected to the lower set of points, ie the left-hand cylinder. These two wires are shrouded in a PVC sleeve and pass from the timing cover through an orifice protected by a rubber grommet to the left of the clutch lever arm. After the engine has been replaced in the frame these wires will be connected at their snap connector terminals to the capacitors (if fitted externally) and the ignition coils.
3 Replace the outer timing cover and secure it with its three crosshead screws. Replace the points cover plate a similarly secure it with its two small screws.
4 Tap the gear lever over the gear change quadrant spindle splines and lock it in position by tightening the small bolt adjacent to the gear lever boss. Replace the kick start lever over its shaft and tap the cotter pin through the lever. Screw on its nut and washer and tighten fully to secure the cotter pin.

36 Replacing the engine unit in the frame

1 Replacement of the engine unit is a two man affair unless a garage hoist is available. Refit the two rear engine plates to the rear of the engine casting by securing them with their associated engine mounting bolts. At this stage only fit the rear engine

31.3 Refit both piston circlips with the help of a small screwdriver. Note protecting cloth beneath piston

31.6 Using piston ring clamps to retain the piston rings

31.7 The pistons are held steady by two parallel bars

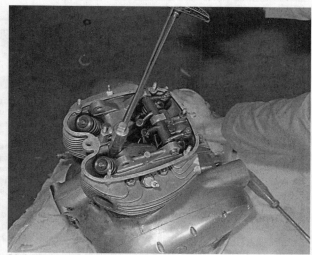

32.1 Tighten the cylinder head bolts with a torque wrench

32.2 Refit exhaust rocker spindle with the shims and thrust
washers located correctly

33.1 Use a feeler gauge to set the tappet clearance

plates loosely.

2 Lift the engine into the frame from the left hand side and position the front mounting lug below the frame lug and the rear mounting above the frame lug, then lever the mountings into position. Insert the mounting studs with washers and nuts through the front lug, the rear lug and finally the bottom lug not forgetting the distance piece which is positioned on the lower engine bolt between the frame and crankcase. Replace the nuts and washers and make all six mounting points absolutely tight.

3 On some models the horn is mounted on a bracket which is bolted to the front engine mounting. It should be refitted at this point.

4 Replace the engine steady stay to the lug on front of the cylinder head and the lug on the base of the steering head. Tighten fully its mounting nuts.

5 Connect the rocker box oil feed pipe to its union at the rear of the cylinder head and then the oil feed and retain pipes to their metal counterparts coming from the base of the engine. Replace the gearbox drain plug and the crankcase drain plug if they have previously been removed.

37.7 Refit the chain link with its open end facing rearwards

37 Engine reassembly - completion and final adjustments

1 Refit the two ignition coils to their mounting bracket which is situated behind the engine and below the front of the seat. Check that both spark plugs are clean and correctly gapped, then refit them. Connect the lead from the upper set of points to the coil serving the right-hand cylinder, and the lead from the lower set of points to the coil serving the left-hand cylinder.

2 Rejoin the electrical connections from the generator, not forgetting those below the crankcase. Replace the fuse in the fuse holder close to the battery, or the battery leads if they were detached.

3 Replace the carburettor(s) using a new gasket between the heat insulating distance piece (if fitted) and the flange of the cylinder head. Insert the slide and needle assembly, making sure the needle locates with the needle jet, before the top of the carburettor(s) if tightened down. Replace the exhaust system by following the reverse of the dismantling procedure.

4 Before replacing the petrol tank, make sure the throttle and air cables are routed so that they will not be trapped or bent into tight covers. Heavy throttle action or a tendency for the slide(s) to stick can often be traced to poor cable routing.

5 The petrol tank is retained by a central bolt fixing and it has an anti-vibration stay at the front secured by a bolt on either side of the tank. Connect the petrol pipes to each petrol tap.

6 Replace the two foot rests and the rear brake pedal and brake rod. The foot rests on early models are mounted on tapers and consequently can be positioned to suit the rider. Note, however, that although the right hand footrest is secured by a nut and washer of normal thread the left hand foot rest is secured by a nut with a left hand thread. Therefore tighten it by turning anti-clockwise. 'Oil-in-frame' frames (1971 to 1973) have foot rests located by pegs and secured by nuts and bolts.

7 Reconnect the final drive chain. It is easiest to insert the spring link if the ends of the chain are pressed into the rear wheel sprocket. Make sure that the closed end of the link faces the direction of chain travel.

8 Refill the oil tank with the recommended grade of engine oil. Refill the gearbox and primary chaincase with the recommended quantity of lubricant.

9 Reconnect the clutch cable and adjust so that there is only a small amount of play at the handlebar lever before the clutch commences to operate. It is possible that the clutch action may be either too heavy or to light. Then, if the clutch springs require adjustment, the primary chaincase and left hand foot rest will have to be removed for the adjustment to be made.

10 Reconnect the speedometer and tachometer drives. Connct the electrical lead to the oil pressure switch at the front of the timing chest (if fitted).

37.9 Adjusting the clutch centre screw

38 Starting and running the rebuilt engine

1 Switch on the ignition and run the engine slowly for the first few minutes, especially if the engine has been rebored. Remove the cap from the top of the oil tank check that oil is returning. There may be some initial delay whilst the pressure builds up and oil circulates throughout the system, but if none appears after the first few minutes running, stop the engine and investigate the cause. If the pressure release valve is unscrewed a few threads, oil should ooze from the joint if the oil pump is building up pressure.

2 Check that all controls function correctly and that the generator is indicating a charge on the ammeter. Check for any oil leaks or blowing gaskets.

3 Before taking the machine on the road, check that all the legal requirements are fulfilled and that items such as the horn, speedometer and lighting equipment are in full working order. Remember that if a number of new parts have been fitted, some running-in will be necessary. If the overhaul has included a rebore, the running-in period must be extended to at least 500 miles, making maximum use of the gearbox so that the engine

runs on a light load. Speeds can be worked up gradually until full performance is obtainable by the time the running-in period is completed.

4 Do not tamper with the exhaust system under the mistaken belief that removal of the baffles or replacement with a quite different type of silencer will give a significant gain in performance. Although a changed exhaust note may give the illusion of greater speed, in a great many cases quite the reverse occurs in practice. It is therefore best to adhere to the manufacturer's specification.

39 Engine modifications and tuning

1 The BSA twins can be tuned to give even higher performance

if desired as can be seen by their success in the past in side-car racing. Many modifications are available both from the manufacturer and various London specialists including 750 cc conversion special, main bearing conversions and racing camshafts.

2 There are several publications, including a pamphlet available from the manufacturer, that provide detailed information about the ways in which a BSA twin engine can be modified to give increased power output. It should be emphasised however that a certain amount of mechanical skill and experience is necessary if an engine is to be developed in this manner and still retain a good standard of mechanical reliability. Often it is perferable to entrust this type of work to an acknowledged specialist and therefore attain the benefit of his experience.

40 Fault diagnosis

Symptom	Cause	Remedy
Engine will not turn over	Clutch slip	Check and adjust clutch.
	Mechanical damage	Check whether valves are operating correctly and dismantle if necessary.
Engine turns over but will not start	No spark at plugs	Remove plugs and check. Check whether battery is discharged.
	No fuel reaching engine	Check fuel system.
	Too much fuel reaching engine	Check fuel system. Remove plugs and turn engine over several times before replacing.
Engine fires but runs unevenly	Ignition and/or fuel system fault	Check systems as though engine will not start.
	Incorrect valve clearances	Check and reset.
	Burnt or sticking valves	Check for loss of compression.
	Blowing cylinder head gasket	See above.
Lack of power	Incorrect ignition timing	Check accuracy of setting.
	Valve timing not correct	Check timing mark alignment on timing pinions.
	Badly worn cylinder barrel and pistons	Fit new rings and pistons after rebore.
High oil consumption	Oil leaks from engine gear unit	Trace source of leak and rectify.
	Worn cylinder bores	See above.
	Worn valve guides	Replace guides.
Excessive mechanical noise	Failure of lubrication system	Stop engine and do not run until fault located and rectified.
	Incorrect valve clearances	Check and re-adjust.
	Worn cylinder barrel (piston slap)	Rebore and fit oversize pistons.
	Worn big end bearings (knock)	Fit new bearing shells.
	Worn main bearings (rumble)	Fit new journal bearings.

Chapter 2 Fuel system and lubrication

Contents

Specifications

Capacities:

Petrol tank:

Lightning, Thunderbolt, Spitfire Mk II and Mk IV, Wasp, Royal Star - To 1968	4 gallons, 4.8 US gallons (18 litres)
All Export models, Hornet and Firebird Scrambler - To 1968	2 gallons, 2.4 US gallons (9 litres)
Lightning, Thunderbolt, Royal Star - 1968 onwards ...	4 gallons, 4.8 US gallons (18 litres) or 3 gallons, 3.6 US gallons (13.5 litres)
Firebird - 1968 onwards	2¾ gallons, 3.3 US gallons (12.5 litres)
All Export models from 1968 on	2½ gallons, 3 US gallons (11.5 litres)

Oil tanks:

All models - All years	6.0 Imp pt/3.4 lit (1962), 5.5 Imp pt/3.1 lit (1963 — 1965), 5.0 Imp pt/2.8 lit (1966 on)

Carburettor: - Models to 1968

	Lightning, Hornet, Firebird Scrambler	Thunderbolt
Type	Amal, Monobloc	
No.	Amal 389/228 r.h. float chamber Amal 689/228 l.h. float chamber	Amal 389/234
Bore	1 5/32 in.	1 1/8 in.
Main jet	270	310
Pilot jet	25	25
Needle jet	.106 in.	.106 in.
Needle position	3	3
Throttle valve	389/3	389/3½
Air cleaner	Dry surgical gauze	Dry surgical gauze

- Models 1966/67

	Mk II Spitfire - 1966/67
Type	Amal, GP2
No.	Amal 516/103 GP r.h. pilot air screw Amal 516/120 GP l.h. pilot air screw
Bore	1 5/32 in.
Main jet	250

Pilot jet	...	...	...	...	...	...	...	...	25
Needle jet	...	...	...	...	...	...	...	...	.109
Needle position	...	...	...	...	...	...	...	...	3
Throttle valve	...	...	...	...	...	...	...	...	5
Air cleaner	...	...	...	...	...	...	...	...	None

- **Models to May 1971**

				Lightning, Hornet Firebird Scrambler	Thunderbolt
Type				Amal, concentric	
No.				Amal R930/21 r.h. concentric Amal L930/22 l.h. concentric	Amal R928/2 r.h. concentric
Bore				30 mm	28 mm
Main jet				190	230
Pilot jet				20	20
Needle jet				.106 in.	.106 in.
Needle position				2	1
Throttle valve				2½	3½
Air cleaner				Dry surgical gauze	Dry surgical gauze

				Hornet/Firebird Scrambler	Spitfire Mk IV
Type				Amal concentric	
No.				Amal R930/24 r.h. concentric Amal L930/25 l.h. concentric	Amal R932/1 r.h. concentric Amal L932/2 l.h. concentric
Bore				30 mm	32 mm
Main jet				190	190
Pilot jet				25	20
Needle jet				.106 in.	.107 in.
Needle position				3	2
Throttle valve				2½	3
Air cleaner				Dry surgical gauze	Dry surgical gauze

- **Models May 1971 to December 1971**

				Lightning	Thunderbolt	Firebird Scrambler
Type				Amal concentric	Amal concentric	Amal concentric
No.				Amal R930/70 r.h. concentric Amal L930/71 l.h. concentric	Amal R918/17 r.h. concentric	Amal R930/72 r.h. concentric Amal L930/73 l.h. concentric
Bore				30 mm	28 mm	30 mm
Main jet				200	230	220
Pilot jet						
Needle jet				.106 in.	.106 in.	.106 in.
Needle position				1	1	1
Throttle valve				3	3½	3
Air cleaner type				Filter cloth	Filter cloth	Filter cloth

- **Models December 1971 onwards**

				Lightning	Thunderbolt	Firebird Scrambler
Type				Amal concentric	Amal concentric	Amal concentric
No.				Amal R930/34 r.h. concentric Amal L930/35 l.h. concentric	Amal R928/6 r.h. conc.	Amal R930/34 r.h. concentric Amal L930/35 l.h. concentric
Bore				30 mm	28 mm	30 mm
Main jet				180	230	180
Pilot jet				622/107	622/107	622/107
Needle jet				.106 in.	.106 in.	.106 in.
Needle position				1	1	1
Throttle valve				3	3½	3
Air cleaner				Filter cloth	Filter cloth	Filter cloth

- **Models**

								A50 Wasp	Royal Star
Type								Amal, monobloc	Amal, monobloc
No.								Amal 389/227 (r.h. float chamber) Amal 689/227 (l.h. float chamber	Amal 376/321
Bore								1 1/8 in.	1 in.
Main jet								200	260
Pilot jet								25	25
Needle jet								.106 in.	.106 in.
Needle position								2	3
Throttle valve								389/3½	376/3½
Air cleaner								Gauze	Gauze

										Wasp	Royal Star
Type ...	...	...	...	...	...	...	...	...	...	Amal, concentric	Amal concentric
No. ...	...	...	...	...	...	...	...	...	...		Amal R626/7 (r.h. concentric)
Bore ...	...	...	...	...	...	...	...	...	...		26 mm
Main jet	...	...	...	...	...	...	...	...	...		200
Pilot jet	...	...	...	...	...	...	...	...	...		25
Needle jet	...	...	...	...	...	...	...	...	...		.106 in.
Needle position	...	...	...	...	...	...	...	...			2
Throttle valve	...	...	...	...	...	...	...	...	...		3½
Air cleaner		...	...	...	...	...	...	...			Dry surgical gauze

1 General description

The fuel system comprises a petrol tank from which petrol is fed by gravity to the float chamber(s) of the carburettor(s). Two petrol taps, with built-in gauze filter, are located one each side beneath the rear end of the petrol tank, For normal running the right hand tap alone should be opened except under high speed and racing conditions. The left hand tap is used to provide a reserve supply, when the main contents of the petrol tank are exhausted.

For cold starting the carburettor(s) incorporate an air slide which acts as a choke controlled from a lever on the handlebars. As soon as the engine has started, the choke can be opened gradually until the engine will accept full air under normal running conditions.

Lubrication is effected by the 'dry sump' principle in which oil from the separate oil tank is delivered by gravity to the mechanical oil pump located within the timing chest. Oil is distributed under pressure from the oil pump through drillings in the crankshaft to the big ends where the oil escapes and is fed by splash to the cylinder walls, ball journal main bearings and the other internal engine parts. Pressure is controlled by a pressure release valve, also within the timing chest. After lubricating the various engine components, the oil falls back into the crankcase, where it is returned to the oil tank by means of the scavenge pump. A bleed-off from the return feed to the oil tank is arranged to lubricate the rocker arms and valve gear, after which it falls by gravity via the pushrod tubes and the tappet blocks, to the crankcase.

2 Petrol tank - removal and replacement

1 Turn off the petrol taps and disconnect the petrol pipes by unscrewing the petrol tank unions.
2 Remove the tank by unscrewing the central tank nut which is beneath the rubber grommet on top of the tank.
3 Remove also the petrol tank anti-vibration bar which is fitted on the underside of the tank at the front. This is secured by two 1/16" bolts.
4 The petrol tank is now free to be lifted clear of the frame.
5 When replacing the tank special care must be taken to ensure that none of the carburettor control cables are trapped or bent to a sharp radius. Apart from making control operation much heavier there is risk that the throttle may stick.

3 Petrol taps - removal and replacement

1 The petrol taps are threaded into inserts in the rear of the petrol tank, at the underside. Neither tap contains provision for turning on a reverse quantity of fuel. It is customary to use the right hand tap only so that the left hand tap will supply the reserve quantity of fuel, unless the machine is used for high speed work or racing. In these latter cases, it is essential to use both taps in order to obviate the risk of fuel starvation.
2 Before either tap can be unscrewed and removed, the petrol tank must be drained. When the taps are removed each gauze filter, which is an integral part of the tap body, will be exposed.
3 When the taps are replaced, each should have a new sealing washer to prevent leakage from the threaded insert in the

bottom of the tank. Do not overtighten; it should be sufficient just to commence compressing the fibre sealing washer.

4 Petrol feed pipes - examination

1 Plastic feed pipe of the transparent variety are used with a union connection to each petrol tap and a push-on fit at the carburettor float chamber.
2 After lengthy service, the pipes will discolour and harden gradually due to the action of the petrol. There is no necessity to renew the pipes at this stage unless cracks become apparent or the pipe becomes rigid and 'brittle'.

5 Carburettor(s) - removal

1 Both single and twin carburettor fitments have been used depending on the version. Early models used the Amal Monobloc carburettor(s) whilst later versions use the Amal Concentric carburettor(s). Both types are described here but special emphasis is given to the concentric because it is, by now, the most usual fitment or replacement.
2 Before removing a carburettor it is first necessary to detach the mixing chamber top which is retained by two small screws and lift away the top complete with the control cables, throttle valve and air slide assemblies. The petrol pipe can then be pulled off the push connection at the float chamber (or the union complete detached) and, after detaching the two retaining nuts and shakeproof washers, the complete carburettor, may be removed from the cylinder head.

6 Carburettor(s) - dismantling, examination and reassembly

Amal concentric carburettor only Fig.2.1

1 To remove the float chamber, unscrew the two crosshead screws on the underside of the mixing chamber. The float chamber can then be pulled away complete with float assembly and sealing gasket. Remove the gasket and lift out the horseshoe-shaped float, float needle and spindle on which the float pivots.
2 When the float chamber has been removed, access is available to the main jet, jet holder and needle jet. The main jet threads into the jet holder and should be removed first, from the underside of the mixing chamber. Next unscrew the jet holder which contains the needle jet. The needle jet cannot be removed until the jet holder has been unscrewed and removed from the mixing chamber because it threads into the jet holder from the top. There is no necessity to remove the throttle stop or air adjusting screws.
3 Check the float needle for wear which will be evident in the form of a ridge worn close to the point. Renew the needle if there is any doubt about its condition, otherwise persistent carburettor flooding may occur.
4 The float itself is unlikely to give trouble unless it is punctured and admits petrol. This type of failure will be self-evident and will necessitate renewal of the float.
5 The pivot needle must be straight - check by rolling the needle on a sheet of plate glass.
6 It is important that the gasket between the float chamber and the mixing chamber is in good condition if a petrol tight joint is

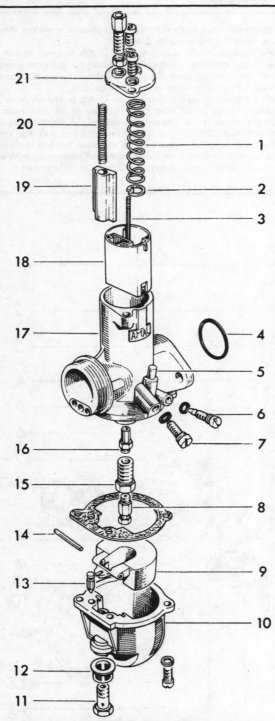

to be made. If it proves necessary to make a replacement gasket, it must follow the exact shape of the original. A portion of the gasket helps retain the float pivot in its correct location; if the pin rides free it may become displaced and allow the float to rise, causing continual flooding and difficulty in tracing the cause. Use Amal replacements whenever possible.

7 Remove the union at the base of the float chamber and check that the inner nylon filter is clean. All sealing washers must be in good condition.

8 Make sure that the float chamber is clean before replacing the float and float needle assembly. The float needle must engage correctly with the lip formed on the float pivot; it has a groove that must engage with the lip. Check that the sealing gasket is placed OVER the float pivot spindle and the spindle is positioned correctly in its seating.

9 Check that the main jet and needle jet are clean and un-obstructed before replacing them in the mixing chamber body. Never use wire or any pointed instrument to clear a blocked jet, otherwise there is risk of enlarging the orifice and changing the carburation. Compressed air provides the best means, using a tyre pump if necessary.

10 Before refitting the float chamber, check that the jet holder and main jet are tight. Do not invert the float chamber, other-wise the inner components will be displaced as the retaining screws are fitted. Each screw should have a spring washer to obviate the risk of slackening.

11 When replacing the carburettor, check that the O-ring seal in the flange mounting is in good condition. It provides an airtight seal between the carburettor flange and the cylinder head flange to ensure the mixture strength is constant. Do not overtighten the carburettor retaining nuts for it is only too easy to bow the flange and give rise to air leaks. A bowed flange can be corrected by removing the O-ring and rubbing down on a sheet of fine emery cloth wrapped around a sheet of plate glass, using a circular motion. A straight edge will show if the flange is level again, when the O-ring can be replaced and the carburettor refitted.

12 Before the mixing chamber top is replaced, check the throttle valve for wear. A worn valve is often responsible for a clicking noise when the throttle is opened and closed. Check that the needle is not bent and that it is held firmly by the clip.

Amal monobloc carburettor only Fig.2.2

Early models were fitted with the Amal monobloc car-burettor which preceded the concentric type, currently in use. Since the two designs of carburettor differ in a number of respects, revised procedure is necessary when dismantling, examining and reassembling the monobloc instrument.

13 The float chamber is an integral part of the monobloc carburettor and cannot be separated. Access is gained by removing three countersunk screws in the side of the float chamber and removing the end cover and gasket. Remove the small brass distance piece and the float needle which will free from its seating as the float is withdrawn.

14 The main jet threads into the main jet holder which itself is screwed into the main body of the mixing chamber. Removal of the lower main jet cover gives access to the main jet. If the hexagonal nut above the jet cover is unscrewed, the main jet holder can be detached and the needle jet unscrewed from the upper end. The pilot jet has its own separate cover nut. When removed, the jet can be unscrewed. It is threaded at one end and has a screwdriver slot.

15 Unless internal blockages are suspected, or the body is worn badly, there is no necessity to remove the jet block, which is a tight push fit within the mixing chamber body. It is removed by pressing upward, through the orifice of the main jet holder, after removing the small locating peg which threads into the car-burettor body. Extreme care must be exercised to prevent distorting either the jet block or the carburettor body which is cast in a zinc-based alloy.

16 Check the float needle for wear by examining it closely. If a ridge has worn around the needle, close to the point, the needle should be discarded and a new one fitted.

FIG. 2.1. COMPONENT PARTS OF THE CONCENTRIC CARBURETTOR

1	Throttle return spring	11	Banjo union bolt
2	Needle clip	12	Filter
3	Needle	13	Float needle
4	'O' ring	14	Float hinge
5	Tickler	15	Jet holder
6	Pilot jet screw	16	Needle jet
7	Throttle stop screw	17	Mixing chamber body
8	Main jet	18	Throttle valve (slide)
9	Float	19	Air slide (choke)
10	Float chamber	20	Air slide return spring
		21	Mixing chamber top

17 The float is unlikely to give trouble unless it is punctured, in which case a replacement is essential. Do not omit to fit the small brass distance piece on the float pivot, after the float has been inserted. If this part is lost, there is nothing to prevent the float moving across to the float chamber end cover and binding - a fault that will give rise to intermittent flooding and prove difficult to pinpoint.

18 There must be a good seal between the float chamber end cover and the float chamber. Always use a new gasket when the seal is broken to obviate the risk of continual petrol leakage.

19 Do not omit to inspect and, if necessary, clean the nylon filter within the float chamber union. When replacing the filter, position it so that the gauze is facing the inflow of petrol. On some of the earlier filters, the plastic dividing strips between the gauze segments are somewhat wide and could impede the flow of petrol under full flow conditions.

20 As stressed in the preceding part of this section, do not use wire on any pointed object to clear blocked jets. Compressed air should be used to clear blockages; even a tyre pump can be utilised if a compressed air line is not available.

21 The monobloc carburettor has an O-ring in the centre of the mounting flange which must be in good condition if air leaks are to be excluded. If the flange is bowed, as the result of previous overtightening, the O-ring should be removed and the flange rubbed down on fine emery cloth wrapped around a sheet of plate glass. Rub with a rotary motion and when a straight edge shows the flange is level again, the O-ring can be replaced.

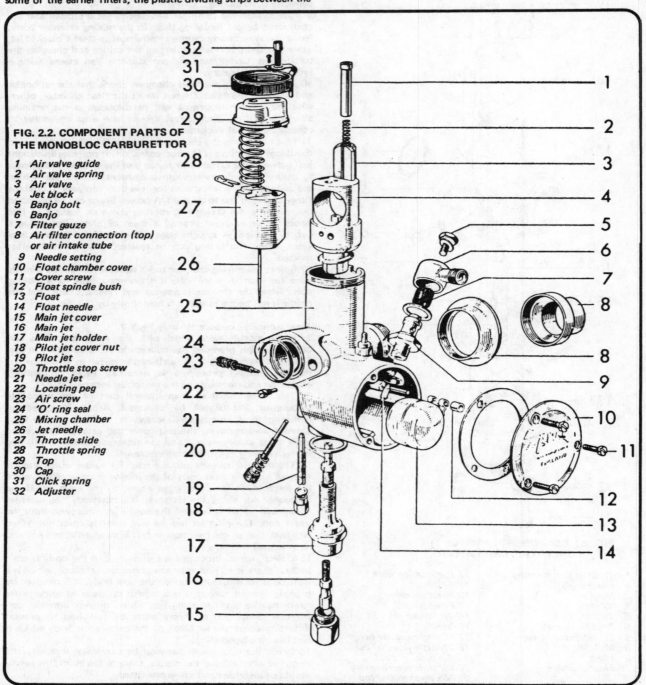

FIG. 2.2. COMPONENT PARTS OF THE MONOBLOC CARBURETTOR

1   Air valve guide
2   Air valve spring
3   Air valve
4   Jet block
5   Banjo bolt
6   Banjo
7   Filter gauze
8   Air filter connection (top) or air intake tube
9   Needle setting
10 Float chamber cover
11 Cover screw
12 Float spindle bush
13 Float
14 Float needle
15 Main jet cover
16 Main jet
17 Main jet holder
18 Pilot jet cover nut
19 Pilot jet
20 Throttle stop screw
21 Needle jet
22 Locating peg
23 Air screw
24 'O' ring seal
25 Mixing chamber
26 Jet needle
27 Throttle slide
28 Throttle spring
29 Top
30 Cap
31 Click spring
32 Adjuster

Amal GP2 racing carburettor only, Fig.2.3

22 The GP2 carburettor has been designed with a view to obtaining the maximum possible power from the engine, at the same time maintaining a progressive and consistent acceleration throughout the throttle range.

23 The float chamber recommended and normally fitted to the GP2 carburettor is a remotely mounted type 510 and is of the bottom feed design incorporating a lever type float.

24 The petrol level in the 510 float chamber is .640 in below the cover joint and is marked with a raised line on the outside of the body. In positioning the float chamber this line should be on a level with the lowest point of the circular scribe mark on the air jet plug (25). Fig. 2.3.

Amal GP2 tuning procedure

25 To obtain accurate carburation the following four operations should be followed closely.

26 A suitable main jet should be selected. The data at the beginning of the Chapter gives the factory specification for running on petrol.

27 The pilot adjustment is achieved when the engine is at its normal running temperature by rotating the pilot air adjuster clockwise to richen the mixture and anti-clockwise to weaken the mixture. Adjust this very gradually until a satisfactory tickover is obtained.

28 Having set the pilot air adjuster open up the throttle progressively and note positions where, if at all, the exhaust note becomes irregular. If this is noticed leave the throttle open at this point and close the air lever slightly; this will indicate whether the spot is rich or weak. If it is a rich spot fit a throttle valve with more cutaway on the air intake side (or vice-versa if it is weak). The above indicates general procedure but the throttle valve size in the specifications is more than adequate for use when running on petrol and should be used.

29 Tuning sequence 26 and 27 will affect the carburation up to somewhere over one quarter throttle after which the jet needle which is suspended from the needle valve comes into action and when the throttle is opened further and tests are again made for rich or weak spots, as previously outlined, the needle can be raised to richen or lowered to weaken the mixture, whichever may be found necessary. With there adjustments correctly made and the mainjet size settled a perfectly progressive mixture will be obtainable from tick over to full throttle.

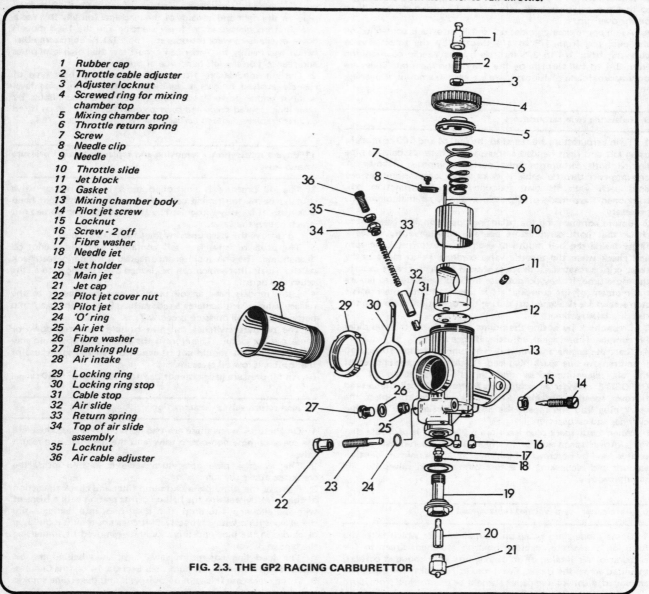

1   Rubber cap
2   Throttle cable adjuster
3   Adjuster locknut
4   Screwed ring for mixing chamber top
5   Mixing chamber top
6   Throttle return spring
7   Screw
8   Needle clip
9   Needle
10 Throttle slide
11 Jet block
12 Gasket
13 Mixing chamber body
14 Pilot jet screw
15 Locknut
16 Screw - 2 off
17 Fibre washer
18 Needle jet
19 Jet holder
20 Main jet
21 Jet cap
22 Pilot jet cover nut
23 Pilot jet
24 'O' ring
25 Air jet
26 Fibre washer
27 Blanking plug
28 Air intake
29 Locking ring
30 Locking ring stop
31 Cable stop
32 Air slide
33 Return spring
34 Top of air slide assembly
35 Locknut
36 Air cable adjuster

FIG. 2.3. THE GP2 RACING CARBURETTOR

7 Carburettor(s) - checking the settings

1   The various sizes of jets and that of the throttle slide, needle and needle jet are predetermined by the manufacture and should not require modification. Check with the specifications list if there is any doubt about the values fitted.

2   Slow running is controlled by a combination of the throttle stop and air regulating screw settings. Commence by screwing the throttle stop screw(s) inward so that the engine runs at a fast tickover speed. Adjust the air screw setting(s) until the tickover is even, without either misfiring or 'hunting'. Screw the throttle stop (screws) outward again until the desired tickover speed is obtained, then recheck with the air regulating screw(s) so that the tickover is as even as possible. Always make these adjustments with the engine at normal running temperature and remember that an engine fitted with high-lift cams is unlikely to run evenly at very low speeds no matter how carefully the adjustments are made.

3   If desired, there is no reason why the throttle stop screw(s) should not be lowered so that the engine will stop completely when the throttle is closed. Some riders prefer this arrangement so that the maximum braking effort of the engine can be utilised on the over-run.

4   As an approximate guide, up to 1/8 throttle is controlled by the pilot jet, from 1/8 to 1/4 throttle by the throttle valve cutaway, from 1/4 to 3/4 throttle by the needle position and from 3/4 to full throttle by the size of the main jet. These are only approximate divisions; there is a certain amount of overlap.

8   Balancing twin carburettors

1   Twin carburettors are fitted to the 650 cc and 500 cc models, using left and right handed carburettors. There is a balance pipe linking both carburettor inlet ports to improve tickover. A one-into-two throttle cable and air slide cable assemblies are used each with its own junction box. The junction box components are made of a plastic material; no maintenance is necessary.

2   Before commencing the balancing operation, it is essential to check that both carburettors operate simultaneously. Place a finger inside the bell mouth of each carburettor intake in turn and check when the throttle valve commences to move as the twist grip is rotated. Both slides should begin to rise at exactly the same time; if they do not, use the cable adjusters to ensure the moment of lift coincides. It is important that the throttle stop screws are slackened off during this operation to obviate the risk of a false reading.

3   Cross-check by noting the points at which the throttle slides lift completely and again, adjusting if necessary.

4   Start the engine and when it is at running temperature, stop it and remove one spark plug lead. Restart the engine and adjust the air regulating screw and throttle stop screw of the OPPOSITE cylinder as detailed in Section 7.2 until the desired tickover speed is obtained. Stop the engine again, replace the spark plug lead and repeat the whole operation with the other cylinder and carburettor.

5   When both spark plug leads are replaced, it is probable that the tickover speed will be too high. It can be reduced to the desired level by unscrewing both throttle stop screws an identical amount and rechecking to ensure both throttle valves still lift simultaneously.

9   Air cleaner - removal and replacement

1   Some models are fitted with an air cleaner which takes the form of a separate, circular housing attached direct to the carburettor air intake, or an oblong box with rounded corners, mounted across the frame. Two types of filter element have been employed, a convoluted paper element or one formed from cloth or felt.

2   None of the filter elements should be soaked in oil. It is sufficient to detach the paper element and blow it clean with compressed air or in the case of the cloth or felt elements, to wash them in paraffin and allow them to drain thoroughly before replacing.

3   On no account run the machine with the air cleaner disconnected unless the carburettor has been re-jetted to suit. When an air cleaner is fitted, it is customary to reduce the size of the carburettor main jet, in order to compensate for the enriching effect of the air cleaner element. In consequence, a permanently-weakened mixture will result if the air cleaner is detached; this will cause failure of the valves and/or piston crown.

10 Exhaust system - general

1   Three separate types of exhaust system are fitted to the 650/500 cc twins, the type depending on the specification of the model concerned. Most machines have downswept exhaust system of the two pipe and two silencer type which may or may not be joined together by a balance pipe close to the exhaust ports. The early Hornets and Firebird scramblers had a high level pipe on the right and left side of the machine. Initially this was a straight through system with no silencing. On the 1968 models however silencers were introduced. The 1971 Firebird scrambler which was mainly for American export saw dual high level pipes and silencers on the left hand side of the machine.

2   Current models are fitted with an entirely new type of silencer evolved to give a significant reduction in noise level without undue power loss. These silencers are recognisable by their long tapered shape and reverse cone end. They can be fitted as direct replacements to earlier versions.

11 Engine lubrication - removing and replacing the oil pressure release valve

1   The oil pressure is controlled by a pre-set release valve situated below the timing cover on the front of the right hand crankcase. It is easily recognised by the large hexagon domed nut which houses the valve.

2   To remove the valve unscrew the large hexagon nut.

3   The pressure valve is a self contained unit and cannot be dismantling. The only maintenance needed is the periodical check on the gauze filter which can be cleaned by brushing with a stiff brush and petrol.

4   Later models have an oil pressure switch adjacent to the release valve which actuates a red warning light on the headlamp body when the oil pressure is below 7 psi.

5   The pressure switches can give trouble and breakdown or cause a short circuit. Therefore if the warning light should stay on persistantly it should not be assumed that the sole cause of the trouble is low oil pressure.

6   An oil pressure gauge cannot be fitted to the A65/A50 range.

12 Non-return valves - examination

1   On the BSA twins there are two non-return valves. These are the scavenge pipe non-return valve and the feed line non-return valve.

2   The scavenge pipe non-return valve is located inside the crankcase above the sump plate.

3   Remove the sump plate and sump filter and check the action of the valve by pushing the ball up off its seating with a piece of wire and allowing it to drop. If it drags back into seating easily the valve is functioning properly. If it does not their is a build up of sludge in the pipe and this should be removed by immersing the pipe in petrol.

4   The feed line non-return valve is situated behind the oil pump. For removal of oil pump see Section 14 of this Chapter.

5   If the crankcase fills with oil overnight and the engine smokes

excessively when started in the morning the feed line valve is suspect.

6 Check that the ball is seating properly and that it is not pitted. Check that the spring is not broken or unduly weak.

13 Engine lubrication - location and examination of oil filters

1 There are two filters in the lubrication system which ensure that the oil is always clean and free from grit or metal particles. These an, the sump filter situated above the crankcase sump plate, and the oil tank filter situated at the base of the oil tank on early models and above the reservoir sump plate on later models which have their oil contained in the frame.

2 The sump filter on early and late models can be removed by unscrewing the four ¼ in BSC nuts which retain the sump plate.

3 Remove the filter from the sump plate and flush it out in petrol.

4 Having cleaned it thoroughly remove any old gasket compound from the filter and crankcase. Check the condition of the sump plate gasket. If it is torn replace it.

5 Replace the filter gasket and sump plate using a proprietry jointing compound sparingly.

6 On earlier models with an oil tank as a non-integral part of the frame gain access by turning the fasteners on the right hand side cover and withdraw the cover.

7 The oil tank filter is screwed into the lower right hand corner of the tank. With a funnel and suitable container for the oil at hand, unscrew the filter and allow the oil to drain.

8 Wash the filter thoroughly in petrol and allow to dry.

9 Replace the clean filter body using a new fibre gasket if necessary.

10 On late machines with the oil contained in the frame the oil 'tank' filter is situated behind the gearbox at the base of the large diameter frame tube. The filter cover is similar to the engine sump plate.

11 Remove the drain plug in the centre of the filter cover plate and allow the oil to drain into a suitable container.

12 Remove the cover plate by unscrewing the four self locking nuts.

13 Remove the filter body and clean thoroughly in petrol.

14 Replace the filter body and cover plate after checking that the gasket is in good condition. Replace the drain plug.

14 Engine lubrication - removal, examination and replacement of oil pump

1 Before the oil pump can be removed the inner and outer timing cores must be removed to gain access to the pump. This is dealt with in Chapter 1.9.

2 Remove oil pump as detailed in Chapter 1.10 taking care not to mislay the non-return ball valve and spring.

3 Dismantle the oil pump body by first removing the circlip at the spindle end with a pair of pliers. The thrust washer and spindle can now be removed.

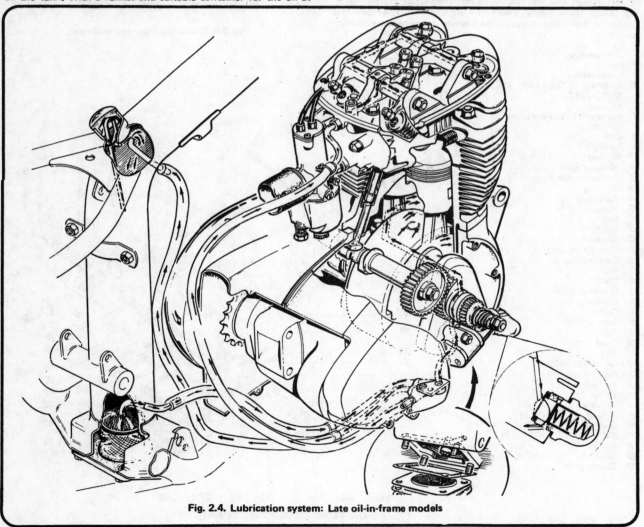

Fig. 2.4. Lubrication system: Late oil-in-frame models

4 Remove the four screws at the base of the pump, the base plate and the spindle housing. The four gears which make up the pump are now exposed.

5 Wash all the oil pump components in petrol and allow them to dry.

6 Examine the components looking for signs of scoring in the pump body and on the base plate. Check also that no foreign matter is lodged between the gear teeth. If excessive scoring is evident the scored parts should be renewed. Slight scoring can be ignored and cleaned up by smoothing with fine emery paper. All traces of emery grit must be removed before reassembly.

7 Wear normally occurs on the spindle teeth. If these are worn to the extent that the sharp edge has gone then the spindle should be renewed.

8 On later models an O-ring is fitted to the feed driving gear to provide a seal between the spindle housing and this gear. The O-ring must be in good condition if it is to be replaced. If in doubt renew it.

9 When inspection has been the pump assembly can be rebuilt. Utter cleanliness must be observed at all times.

10 Insert the feed driving gear wheel into its housing. Place in position the driven feed gear wheel.

11 Insert the return gear wheels on the lower side of the pump.

12 Slide the spindle through the two driven gears and apply clean oil to all four gears.

13 Refit the spindle housing and base plate.

14 Place the driving spindle (the spindle with the worm gear on it) in position and test the pump for freedom of movement. If the pump turns freely insert the spindle thrust washer and circlip. If the pump does not turn freely it has been assembled wrongly or the base plate is scored and is fouling the gear wheels.

15 Refit the pump assembly to the crankcase and secure the three nuts with the torque setting of 7 ft lb. If this procedure is not observed the pump body may distort and prevent the pump from working.

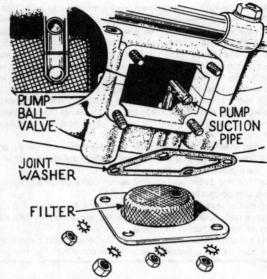

Fig. 2.5. Crankcase filter and ball valve

1 Felt washer
2 Nut
3 Lockwasher
4 Sprocket (17T), (18T), (19T), (20T), (21T)
5 'O' ring
6 Revolution counter drive
7 'O' ring
8 Ball
9 Spring
10 'O' ring
11 Release valve plug
12 Release valve body
13 'O' ring
14 Feed gear
15 Washer
16 Circlip
17 Washer
18 Spindle
19 Spindle housing
20 Lockwasher
21 Nut
22 Stud
23 Scavenge pipe
24 Gasket
25 Cover
26 Lockwasher
27 Nut
28 Stud
29 Nut
30 Oil pump body
31 Driving gear
32 End plate
33 Screw
34 Spindle
35 Driven gear
36 Gasket
37 Ball
38 Spring
39 Stud
40 Driven gear

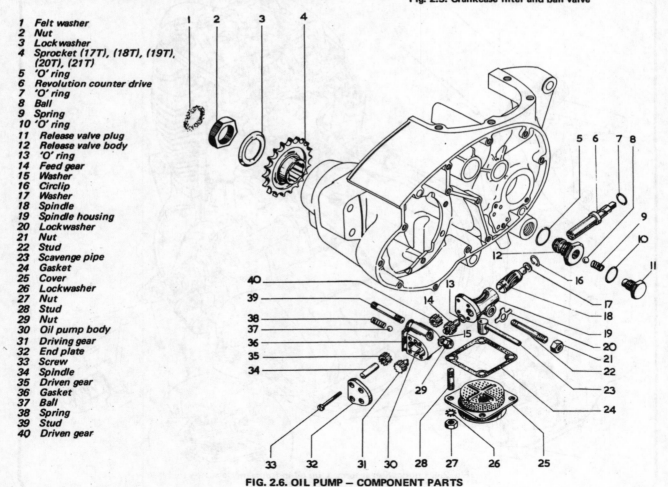

FIG. 2.6. OIL PUMP – COMPONENT PARTS

15 Fault diagnosis

Symptom	Cause	Remedy
Excessive fuel consumption	Air filter choked, damp or oily	Check. Renew if necessary.
	Fuel leaking from carburettor	Check all unions and gaskets.
	Float needle sticking	Float needle seat needs cleaning.
	Worn carburettor	Renew.
Idling speed fast	Throttle stop screw in too far	Readjust screw.
	Carburettor top loose	Tighten top.
Poor throttle response	Rich mixture	Check for displaced or punctured float.
Engine dies after running for a short shile	Blocked air vent in filler cap	Clean.
	Dirt or water in carburettor	Remove and clean float chamber.
General lack of performance	Weak mixture. Float needle stuck	Remove float chamber and check.
	Leak between carburettor and cylinder head	Bowed flange. Rub down until flat and replace 'O' ring seal.
	Fuel starvation	Turn on both petrol taps. Check that petrol filter is not blocked.
Low oil pressure	Insufficient oil in tank	Replenish.
	Blocked filters	Check. Wash in petrol.
	Worn oil pump or big ends	Renew.
	Oil pipes connected wrong way round	Check connections carefully.
	Worn crankshaft timing side bush	Overhaul crankshaft and main bearings

Chapter 3 Ignition system

Contents

Specifications

Ignition coils:

Make	Lucas
Type	MA12 all models except 3ET on Hornet, Firebird and Wasp
Voltage	12 volt

Contact breakers:

Make	Lucas
Type	4 CA to 1968, 6 CA 1968 onwards, 4 CA ET on Hornet, Firebird and Wasp
Gap	0.015 in. all models

Spark plugs:

Make ... ...

Type:

	Champion	NGK	Lodge	KLG
All models 1962 to 1965, all single-carb engines 1966 on ...	N4	B7ES	HLN	FE80
All twin-carb (sports) engines 1966 on	N3	B8ES	2HLN	FE100
Gap	0.020 — 0.025 in (0.51 — 0.64 mm)			

Ignition timing — fully advanced

	Crankshaft angle	Piston position
All models with battery and coil ignition	34° BTDC	0.3045 in (7.7343 mm)
A50 Wasp 1965 to 1966	23° BTDC	0.1427 in (3.6246 mm)
A65 Hornet 1965 to 1967, Firebird 1967 to 1970	28° BTDC	0.2094 in (5.3188 mm)

Note: Models manufactured before 1966 have a 6 volt ignition system (12 volt equipment optional from 1963 on)

1 General description

The spark necessary to ignite the petrol/air mixture in each combustion chamber is derived from a battery and coil, used in conjunction with a contact breaker to determine the precise moment at which the spark will occur. As the points separate the circuit Is broken and a high tension voltage is developed across the points of the spark plug which jumps the air gap and ignites the mixture. Each cylinder has its own ignition circuit, hence the need for two separate ignition coils and a twin contact breaker assembly.

When the engine is running, the surplus current produced by the generator is converted into direct current by the rectifier and used to charge the battery. The generator provides sufficient current for the initial start even when the battery is fully discharged. Generator output does not correspond directly to engine rpm and is regulated by a diode in circuit, eliminating the need for an electro-mechanical device such as a voltage regulator. All coils in the system are brought into operation only if there is a heavy electrical load when all lights are in use.

2 Checking generator output

Specialised test equipment of the multi-meter type is essential to check generator output with any accuracy. It is unlikely that the average owner will have access to this type of equipment or instruction in its use. In consequence, if generator performance is suspect, it should be checked by a Triumph agent or an auto-electrical expert.

3 Ignition coils - checking

1 Ignition coil is a sealed unit, designed to give long service without need of attention. A twin coil system is used on the BSA unit-construction twins, with the coils mounted on the tube that joins the duplex seat tubes. It is necessary to remove both side covers to gain access.

2 To check whether a coil is defective, disconnect the spark plug lead from the plug concerned and turn the engine over until the contact breaker points that relate to the coil being tested are closed (check with colour coding of wire). Switch on the ignition and hold the plug lead about 3/16 in away from the cylinder head. If the coil is in good order, a strong spark should jump the air gap between the end of the plug lead and the cylinder head, each time the points are flicked open.

3 A coil is most likely to fail if the outer casing is compressed or damaged in any way. Fine gauge wire is used for the secondary winding and this will break easily if subjected to any strain by a damaged casing.

4 Contact breakers - adjustment

1 The contact breaker points are located behind the cover attached to the forward end of the timing cover by two cross-head screws. Remove both screws and withdraw the cover.

2 Two types of contact breaker assembly have been used. Machines prior to 1968 have the Lucas 4CA type containing condensers. Later models utilise the Lucas 6CA assembly, which is more accessible because the condensers have been transferred to another location.

3 In each case, the correct contact breaker gap is 0.015 in with the points open fully. To adjust the gap on the 4CA unit, slacken the slotted nut that secures the stationary contact point and move the contact either inwards or outwards until the gap is correct. Tighten the nut and recheck that the setting is still correct. Repeat this procedure for the other set of points.

4 The 6CA unit has the fixed contact point secured by a locking screw which must be slackened first. Adjustment is effected by turning an eccentric screw in the forked end of each contact breaker point assembly, before tightening the locking screw. Recheck that the setting is correct before repeating the procedure for the second set of points. With the 6CA unit, checking whether the points are open fully is simplified by aligning a scribe mark on the end of the contact breaker cam with the nylon heel of the points set, in each case.

5 It is sometimes found that there is a discrepancy between the points gaps of the 6CA unit when the scribe mark of the cam is aligned with the nylon heels. If the discrepancy is greater than 0.003 it is probably caused by cam run-out and can be cured by tapping the cam with a soft metal drift until it seats correctly. Cases have also occurred where the edge of one of the secondary backplates has fouled the cam. Contact between the cam and the backplate can result in the automatic advance unit remaining in the permanently retarded position, so if run-out is evident, either of these two faults should be investigated and remedied.

5 Contact breaker points - removal, renovation and replacement

1 If the contact points are burned, pitted or worn, they must be removed for dressing. If, however, it is necessary to remove a substantial amount of material before the faces can be restored, the points should be renewed.

2 To remove the contact breaker points from the 4CA unit, remove the securing nuts from the condenser terminals. This will free the return spring from the moving contacts, which can be withdrawn from their respective pivot pins. Removal of the slotted nuts will free the fixed contact points.

3 In the case of the 6CA unit, the moving points are removed by unscrewing the nut that secures each low tension lead wire

and removing the lead and nylon brush. The return spring and contact point can then be withdrawn from each pivot pin. The fixed points are each retained by two screws which secure them to the backplate assembly.

4 The points should be dressed with an oilstone or fine emery cloth, taking care to keep them absolutely square throughout the dressing operation. If this precaution is not observed, the points will make angular contact with one another when they are replaced, and will burn away rapidly.

5 Replace the points by reversing the dismantling procedure. Take particular care to replace the insulating washers and spacers in their original positions, otherwise the points will be isolated electrically and the ignition system will no longer function.

6 It is important that the points are maintained in a clean conditon.

7 The timing side (right hand) cylinder ignition system is operated by the upper set of points, to which is attached the black/yellow wire (early models, 1969 on wiring may differ).

6 Condensers - removal and replacement

1 As mentioned in the preceding section, the condensers used in conjunction with the Lucas 4CA contact breakers assembly are contained within the contact breakers housing, attached to the backplate. It is advisable to withdraw the assembly complete before detaching the condensers, by removing the two pillar bolts securing the backplate to the timing cover housing.

2 When the Lucas 6CA contact breaker assembly is fitted, the condensers are located remotely. They are attached to a plate beneath the seat.

3 If the engine becomes difficult to start, or if misfiring occurs, it is probable that a condenser has failed. Note that it is rare for both condensers to fail simultaneously, unless they have been damaged in an accident. Examine the contact breaker points whilst the engine is running to see whether arcing is taken place and when the engine is stopped, examine the faces of the points. Arcing taking place or the points having a blackened and burnt appearance is characteristic of condenser failure.

4 It is not possible to check the condenser without the necessary test equipment. It is therefore best to fit a replacement condenser and observe the effect on engine performance, especially in view of the low cost of the replacement.

7 Ignition timing - checking and resetting - 1965 to 1968 models

There are two main methods available for securing accurate ignition timing. These are, the static method which involves locking the engine in a prescribed position on the compression stroke and adjusting the points until they are just open, and the stroboscopic method which involves specialised equipment, and is really an extension of the static timing method for obtaining dead accurate ignition timing.

1 If the static timing method is selected, commence by removing the rocker box cover and both spark plugs.

2 Remove the contact breaker points cover by unscrewing the two crosshead screws securing the cover plate at the front of the timing case.

3 Regardless of at what position the engine is timed, using a feeler gauge set both points gaps to 0.015" when fully open. See Section 4 of this Chapter.

4 To locate the crankshaft and pistons in the fully advanced position it is first necessary to remove the crankcase cover plate, shown in Fig.3.3.

5 There is a plug supplied in the tool kit for locking the crankshaft in the fully advanced position. This plug is inserted in the crankcase aperture and the crankshaft slowly turned until the plug locates with the groove in the crankshaft.

6 The crankshaft can either be turned using the kickstart lever or if the engine is still in the frame 4th gear can be selected and the back wheel slowly turned until the plug and groove line up.

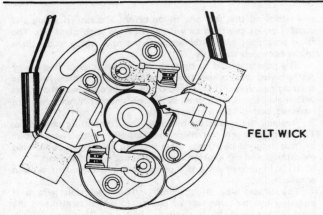

FELT WICK

Fig. 3.1. The Lucas 4CA contact breaker

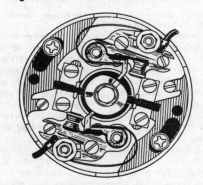

Fig. 3.2. The Lucas 6CA contact breaker

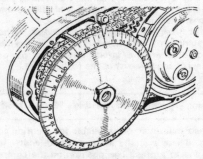

1962 to 1964 models - using a degree disc

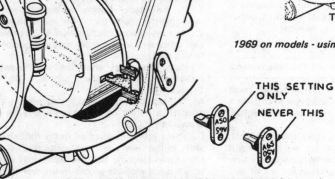

1965 to 1968 models - using BSA tool 68-710 (see text)

Do not use force.

7   Do not attempt to remove the crankshaft whilst the timing plug is located in position.

8   Having located the flywheel by means of the plug, check which cylinder is on the compression stroke by examining the valves. If both valves are closed on the right hand cylinder then the upper set of contacts should be used. For the left hand cylinder use the lower set.

9   It is important when timing to know exactly when the points are just opening. An easy and simple method is to connect a battery and bulb across the points being checked. One side of the points is earthed so one lead can be secured to the crankcase at some suitable point and the other connection should be made with a small crocodile clip on the 'C' spring of the contact breaker being checked. When the points are closed the bulb will light up and when they are just opening it will go out..

10   If, say the right hand cylinder is on its compression stroke, connect the crocodile clip to the C spring of the upper set of points, remembering to have one lead from the bulb secured and earthed to the crankcase.

11   Loosen the contact breaker backplate screws.

12   Hold the cam and turn it anti-clockwise so that the bob weights of the advance-retard mechanism are in the fully open position.

13   Holding the cam in this position turn the contact breaker plate so that the upper set of points are just opening ie the bulb should go out just as the points are opening.

14   Lock the plate in position by tightening its two retaining screws and recheck the setting.

15   Take out the timing plug, turn the engine through one revolution and reinsert the timing plug.

16   Place the crocodile clip on the 'C' spring of the lower set of points. Turn the cam to the fully advanced position.

17   For the lower set of points do not move the contact breaker back plate to obtain the timing but adjust the points so that they are just opening, once again using the bulb and battery to obtain accuracy.

18   On later models there is provision for stroboscopic timing. This is the most accurate method available.

19   Initially time the engine by the timing plug method if the timing is known to be way out.

20   Locate the cover on the front side of the primary chaincase and remove its four retaining screws and the cover.

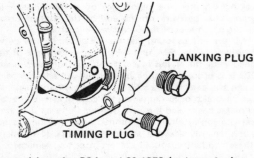

BLANKING PLUG

TIMING PLUG

1969 on models - using BSA tool 60-1859 (early version)

THIS SETTING ONLY

NEVER THIS

Fig. 3.3 Static ignition timing check - positioning the crankshaft

21 The aperture will reveal the alternator rotor with two marks on it 180° apart and a pointer set in the chaincase at the base of the aperture.

22 Using an independent battery source for the stroboscope connect one lead to the right hand spark plug and the other lead grounded to the motorcycle.

23 Start the engine and rev up to 3000 rpm.

24 Shine the strobe light on to the rotor, which will appear stationary. If the mark on the rotor does not line up with the fixed pointer, align by rotating the contact breaker backing plate appropriately (left to retard, right to advance).

25 Switch the stroboscope lead from the right hand spark plug to the left hand plug and repeat. To alter the timing of the left hand cylinder alter the points gap of the lower set of points. Open them to retard, close them to advance.

26 Having timed the engine remove the stroboscope and replace the primary chaincase inspection cover.

8 Automatic advance/retard unit - removal, examination and replacement

1 Fixed ignition timing is of little advantage as the engine speed increases and it is therefore necessary to incorporate a method of advancing the timing by centrifugal means. A balance weight assembly located behind the contact breaker, linked to the contact breaker cam is employed in the case of the BSA unit construction twins. It is secured to the idler wheel by a bolt that passes through the centre of the contact breaker cam. It can be withdrawn after the outer timing cover and contact breaker assembly has been removed.

2 The unit is most likely to malfunction as the result of condensation, which will cause rusting to take place. This will immediately be evident when the assembly is removed. Check that the balance weights move quite freely and that the return springs are in good order. Before replacing the assembly by reversing the dismantling procedure, lubricate the balance weight pivot pins and the cam spindle, amd place a slight smear of grease on the face on the contact breaker cam. Lubricate the felt pad that bears on the contact breaker cam.

9 Ignition cut-out

1 The ignition circuit is controlled by the ignition switch mounted on the left hand side of the steering head on earlier models and on the right hand side cover on later models. The switch is operated by a Yale type ignition key which renders the motorcycle thief proof when parked. The key can only be withdrawn from the switch when it is in the 'OFF' position and the ignition circuit is broken.

2 A state of emergency can occur (perhaps a sticking throttle) when the machine is on the move and it is not convenient to reach out for the switch key. For this reason late models have an additional cut-out button on the right hand side of the handlebars, which will break the ignition circuit all the time it is depressed. Because the ignition circuit is broken only when the cut-out button is depressed, it is essential to turn the switch to the off position and remove the key when the machine is parked.

10 Spark plugs - checking and resetting the gap

1 A 14 mm spark plug is fitted to both cylinders of the BSA unit-construction twins, the grade depending on the model designation. Refer to the Specifications of this Chapter for the recommended grades.

2 All models use spark plugs with a ¾ in reach which should be gapped at 0.025 in. Always use the grade of plug recommended or the exact equivalent in another manufacturer's range.

3 Check the gap at the plug points every 3000 miles or during every six monthly service, whichever is soonest. To reset the gap bend the outer electrode to bring it closer to the inner electrode and check that a 0.025 in feeler gauge can just be inserted. Never bend the centre electrode otherwise the insulator will crack, causing engine damage if particles fall in whilst the engine is running.

4 With some experience, the condition of the spark plug electrodes and insulators can be used as a reliable guide to engine operating conditions. See the accompanying illustrations for examples.

5 Always carry a pair of spark plugs of the correct grade. In the rare event of plug failures, they will enable the engine to be restarted.

6 Never overtighten a spark plug, otherwise there is risk of stripping the threads from the cylinder head, particularly those cast in light alloy. The plugs should be sufficiently tight to seat firmly on their sealing washers. Use a spanner that is a good fit, otherwise the spanner may slip and break the insulator.

7 Make sure the plug caps are a good fit and free from cracks. These caps contain the suppressor that eliminates radio and TV interference.

11 Fault diagnosis

Symptom	Cause	Remedy
Engine will not start	No spark at plug	Check whether contact breaker points are opening and also whether they are clean. Check wiring for break or short circuit.
Engine fires on one cylinder	No spark at plug or defective cylinder	Check as above, then test ignition coil. If no spark, see whether points arc when separated. If so, renew condenser.
Engine starts but lacks power	Automatic advance unit stuck or damaged	Check unit for freedom of action and broken springs.
	Ignition timing retarded	Verify accuracy of timing. Check whether points gaps have closed.
Engine starts but runs erratically	Ignition timing too far advanced	Verify accuracy of timing. Points gaps too great.
	Spark plugs too hard	Fit lower grade of plugs and retest.

Fig. 3.4a. Spark plug adjustment

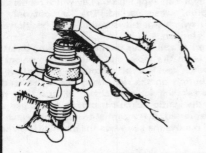

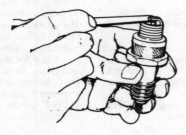

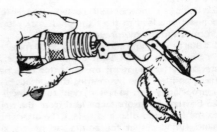

*Cleaning deposits from elec
trodes and surrounding area
using a fine wire brush*

*Checking plug gap with feeler
gauges*

*Altering the plug gap. Note use
of correct tool*

Fig. 3.4b. Spark plug electrode conditions

*White deposits and damaged
porcelain insulation indicating
overheating*

*Broken porcelain insulation
due to bent central electrode*

*Electrodes burnt away due to
wrong heat value or chronic
pre-ignition (pinking).*

*Excessive black deposits
caused by over-rich mixture
or wrong heat value*

*Mild white deposits and elec-
trode burnt indicating too
weak a fuel mixture*

*Plug in sound condition with
light greyish brown deposits*

Chapter 4 Frame and forks

Contents

Specifications

Telescopic forks:

Oil contents per leg ... ...   1/3 pint   190 cc all models

Oil viscosity ... ...   SAE 10W/30 or 10W/40 or SAE 20 engine oil (1962 to 1970),
TQF (1971 on)

1 General description

A full duplex cradle frame is fitted to the BSA construction twins. Early models have a central spine running from the steering head to the rear frame of 15/8" diameter steel tubing. Later models have a large diameter spine from the steering head to the rear frame and downwards to the swinging arm point. This gives increased rigidity and also acts as an oil reservoir contained within the frame.

Rear suspension is provided by a swinging arm assembly that pivots on a plate welded to the rear of the frame cradle together with a lug welded on the central downtube of the large diameter spine, on later models. Movement is controlled by two hydraulically-clamped rear suspension units, one on each side of the rear frame. The units have three-rate adjustment, so that the spring loading can be varied to match the conditions under which the machine is to be used.

Front suspension is provided by telescopic forks of conventional design.

As mentioned earlier, extensive redesigning took place during 1970 and from the 1971 models onwards a different frame of the duplex type was substituted. The most distinctive feature was the use of an enlarged vertical seat pillar to contain the engine oil in place of the hitherto side-mounted oil tank. Front suspension was also changed. Forks of the slimline type replaced the originals having slim, exposed stanchions with internal springs and a different damper assembly.

2 Front forks - removal from frame

1 It is unlikely that the front forks will need to be removed from the frame as a complete unit unless the steering head bearings require attention or the forks are damaged in an accident.

2 Commence operations by placing the machine on the centre stand and disconnecting the front brake. If the nut and bolt through the U shaped connection to the brake operating arm are removed the cable can be unscrewed clear of the cable adjuster lug and removed from the cable guide attached to the front mudguard.

3 To remove the front wheel, unscrew the brake anchor arm nut from the cover plate and slacken off the nuts at the other end to enable the removal of the strap from the plate. Models fitted with a twin leading shoe brake or conical front hub have no brake anchor plate as such and so this arm is omitted.

4 Slacken off the pinch bolt on the left hand fork end and using a bar through the head of the spindle unscrew the spindle in a clockwise direction (left hand thread) and withdraw.

On later models remove the fork end caps by unscrewing the four nuts and removing the same together with their washers.

5 On the earlier models support the wheel as the spindle is withdrawn and when it is clear the wheel can be pulled away from the right hand leg and clear of the machine. On later models the wheel will drop clear after the fork end caps have been removed.

6 It is convenient at this stage to drain the fork legs of their oil

content, if the forks are to be dismantled at a later stage. The drain plug is found above and behind the wheel spindle on both fork legs. Remove both drain plugs and leave the forks to drain into some suitable receptacle whilst the dismantling continues.

7   There is no necessity to remove the front mudguard unless the fork legs are to be dismantled. The lower mudguard stays bolt direct to lugs at the lower end of each fork leg; the centre fixing is made to the inside of each fork leg where a shaped lug accepts the cut out of the mudguard stay assembly which is then retained by a bolt and a washer. If the various nuts, bolts and washers are removed the mudguard can be withdrawn complete with stays.

8   Detach the controls from the handlebars by either disconnecting the cable ends or by detaching the controls with the cables still attached. This includes the cut-out button, horn button and dip switch. In the case of the horn button, it is preferable to detach the main lead from the battery before the button is removed, to prevent short circuits. Detach the speedometer drive cable by unscrewing the gland nut from the bottom of the speedometer head and the pilot bulb.

9   Remove the handlebars by withdrawing the bolts that retain the split mounting clamps to the fork top yoke. Detach rev counter drive cable similarly, if fitted.

10   Remove the headlamp by unscrewing the two retaining bolts in each side of the shell. The headlamp can be left to hang in a position where it is not liable to suffer damage.

11   Unscrew and remove the two chromium plated caps at the top of each fork leg. Slacken the pinch bolt in the fork top yoke, immediately to the rear of the steering head. This will expose the slotted adjusting sleeve, which should be unscrewed (right-hand thread) using either a strip of metal to engage with the slots or the BSA Service Tool designed for this purpose. If the machine is fitted with a steering damper, it will be necessary first to remove the split pin from the bottom end of the stem and unscrew the knob until it can be detached, complete with rod.

12   When the adjusting sleeve has been withdrawn, the fork top yoke can be removed by striking it from the underside with a rawhide mallet, first one side and then the other. The forks, complete with steering head stem can now be drawn downwards until they are completely separate from the machine. Note the uncaged ball bearings of the steering head assembly will drop free as the cups and cones separate, necessitating some arrangement for catching them.

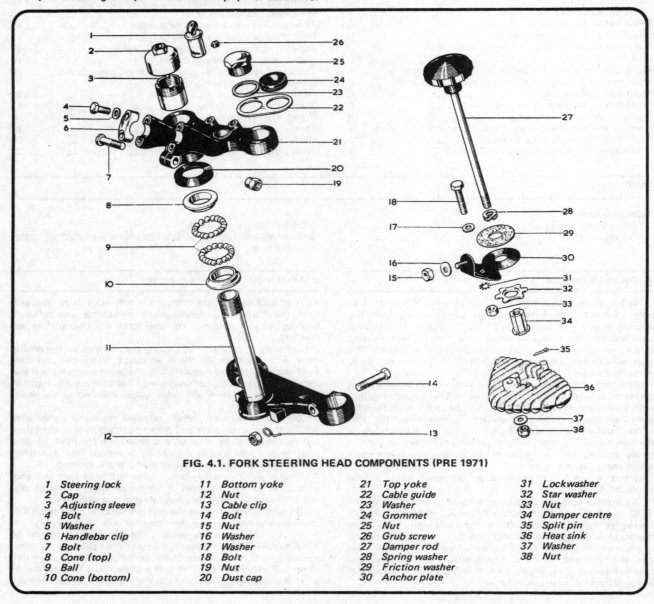

FIG. 4.1. FORK STEERING HEAD COMPONENTS (PRE 1971)

1 Steering lock	11 Bottom yoke	21 Top yoke	31 Lockwasher
2 Cap	12 Nut	22 Cable guide	32 Star washer
3 Adjusting sleeve	13 Cable clip	23 Washer	33 Nut
4 Bolt	14 Bolt	24 Grommet	34 Damper centre
5 Washer	15 Nut	25 Nut	35 Split pin
6 Handlebar clip	16 Washer	26 Grub screw	36 Heat sink
7 Bolt	17 Washer	27 Damper rod	37 Washer
8 Cone (top)	18 Bolt	28 Spring washer	38 Nut
9 Ball	19 Nut	29 Friction washer	
10 Cone (bottom)	20 Dust cap	30 Anchor plate	

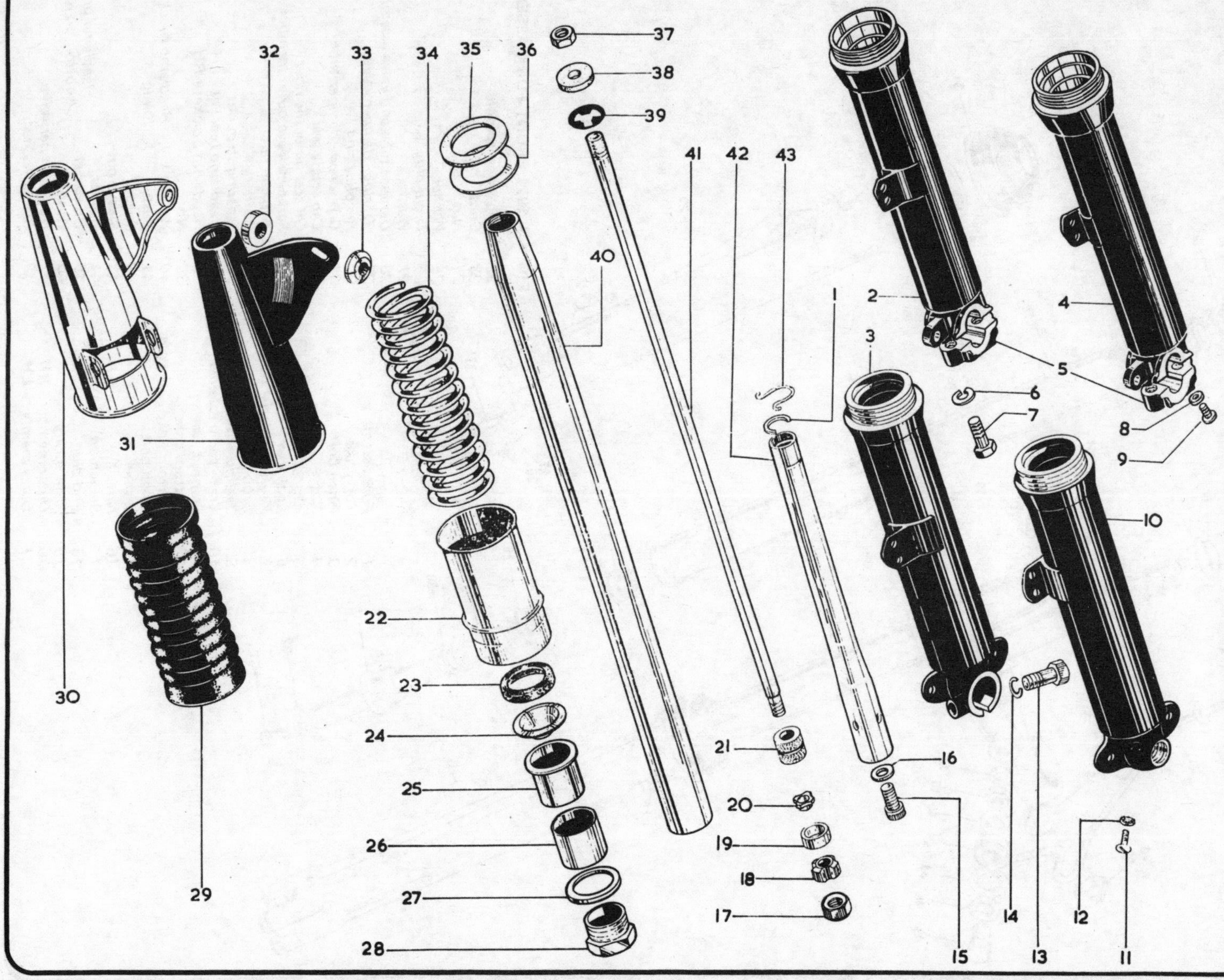

FIG. 4.2. FORK LEG COMPONENTS (PRE 1971)

1 Circlip
2 Sliding tube (left-hand)
3 Sliding tube (left-hand)
4 Sliding tube (right-hand)
5 End cap
6 Spring washer
7 Bolt
8 Fibre washer
9 Plug
10 Sliding tube (right-hand)
11 Plug
12 Fibre washer
13 Bolt
14 Spring washer
15 'Allen' screw
16 Washer
17 Nut
18 Valve seat
19 Damper valve
20 Valve collar
21 Bush
22 Oil seal holder
23 Oil seal
24 Spacer
25 Top bush
26 Bottom bush
27 Washer
28 Plug
29 Bellows
30 Fork sleeve (right-hand)
30 Fork sleeve (left-hand)
31 Fork sleeve (right-hand)
31 Fork sleeve (left-hand)
32 Spacer
33 Washer
34 Spring (solo)
34 Spring (sidecar)
35 Sealing washer
36 Retaining washer
37 Nut
38 Rubber washer
39 Retaining washer
40 Shaft
41 Damper rod
42 Damper tube
43 Circlip

54

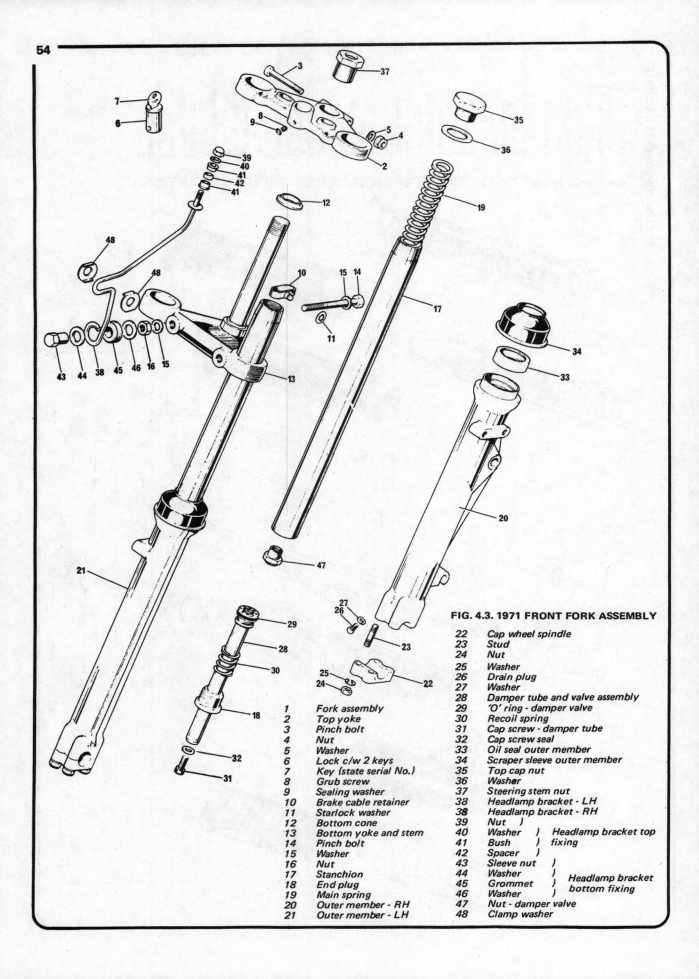

FIG. 4.3. 1971 FRONT FORK ASSEMBLY

1	Fork assembly
2	Top yoke
3	Pinch bolt
4	Nut
5	Washer
6	Lock c/w 2 keys
7	Key (state serial No.)
8	Grub screw
9	Sealing washer
10	Brake cable retainer
11	Starlock washer
12	Bottom cone
13	Bottom yoke and stem
14	Pinch bolt
15	Washer
16	Nut
17	Stanchion
18	End plug
19	Main spring
20	Outer member - RH
21	Outer member - LH
22	Cap wheel spindle
23	Stud
24	Nut
25	Washer
26	Drain plug
27	Washer
28	Damper tube and valve assembly
29	'O' ring - damper valve
30	Recoil spring
31	Cap screw - damper tube
32	Cap screw seal
33	Oil seal outer member
34	Scraper sleeve outer member
35	Top cap nut
36	Washer
37	Steering stem nut
38	Headlamp bracket - LH
38	Headlamp bracket - RH
39	Nut)
40	Washer) Headlamp bracket top
41	Bush) fixing
42	Spacer)
43	Sleeve nut)
44	Washer) Headlamp bracket
45	Grommet) bottom fixing
46	Washer)
47	Nut - damper valve
48	Clamp washer

2.2 Disconnect brake cable from operating arm

2.4a Slacken off the wheel spindle pinch bolt (early models)

2.4b Remove split clamps from bottom of fork legs (later models)

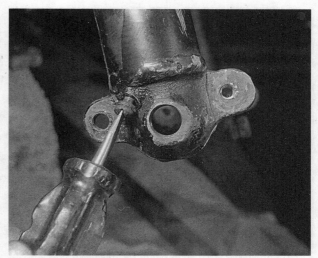

2.6 Remove fork drain plug

2.11a Unscrew the plated caps at the top of each fork leg

2.11b Slacken the pinch bolt in the fork top yoke

2.11c Steering damper must be removed to gain access to ...

3 Front forks - dismantling

1 To remove the individual fork legs, unscrew the pinch bolts in the fork bottom yoke, and unscrew the clips securing the rubber gaiters. The fork legs should now pull clear. If they are still a tight fit, spring open each pinch bolt a little, to ease the grip.

2 Unscrew the chromium plated cover immediately above the lower fork leg. This has a right hand thread and may require the use of a strap wrench or BSA Service Tool No. 61-3005 to unscrew it. When the cover has been removed, it will expose a circlip within the lower fork leg that acts as a limit stop. If the circlip is removed, the inner fork leg can be withdrawn completely, together with the fork bushes.

3 It is not necessary to disturb the head race assembly or to detach the fork yokes from the frame if the fork legs alone are to be dismantled. The legs can be dismantled individually, as described, if the front wheel and mudguard are removed.

2.11d ... unscrew plated cap at top of steering head

3.1 Pull fork legs downward, to release from fork yokes

2.11e Slotted adjuster can be unscrewed using metal strip

3.3a Unscrew plated cover to gain access to circlip

3.3b Circlip is located within lower fork leg and must be prized out

3.3c Inner fork leg will pull out, complete with bushes

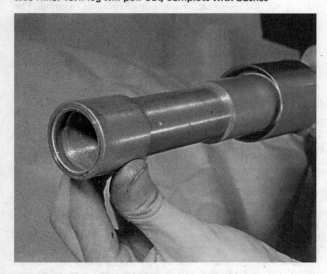

6.1 The dismantled assembly, showing the two fork bushes

4 Front forks - general examination

1 Apart from the oil seals and bushes, it is unlikely that the forks will require any additional attention, unless the fork springs are weak or have to be used with a sidecar attached. If the fork legs or yokes have been damaged in an accident, it is preferable to have them replaced. Repairs are seldom practicable without the appropriate repair equipment and jigs, furthermore there is also a risk of fatigue failure.
2 Visual examination will show whether either the fork legs or the yokes are bent or distorted. The best check for the fork legs is to remove the fork bushes, as described in Section 6 of this Chapter and roll the legs on a sheet of plate glass. Any deviation from parallel will immediately be obvious.

5 Front forks - examination and replacement of oil seals

1 If the fork legs have shown a tendancy to leak oil or if there is any other reason to suspect the condition of the oil seals, now is the time to replace them.
2 The oil seals are retained within the plated covers that thread on to the bottom legs of the forks. A BSA Service Tool No. 61-3006 is available for extracting the oil seals but since the seals have to be replaced there is no reason why they should not be drifted out of position through the slots in the plated covers it follows that the seals should not be disturbed unless replacement is necessary.

6 Front forks - examination and replacement of bushes

1 Some indication of the extent of wear of the fork bushes can be gained when the forks are being dismantled. Pull each fork inner tube out until it reaches the limit of its extension and check the side play. In this position the two fork bushes are closest together, which will show the amount of play to its maximum. Only a small amount of play that is just perceptible can be tolerated. If the play is greater than this, the bushes are due for replacement.
2 It is possible to check for play in the bushes whilst the forks are still attached to the machine. If the front wheel is gripped between the knees and the handlebars rocked to and fro, the amount of wear will be magnified by the leverage at the handlebar ends. Cross-check by applying the front brake and pushing and pulling the front wheel backwards and forwards. It is important not to confuse any play that is evident with slackness in the steering head bearings, which should be taken up first.
3 The fork bushes can be slid off the fork tubes if the tubes are clamped in a vice fitted with soft clamps and the large nut on the extreme end removed. If the replacement bushes are a slack fit on the tubes, wear has occured on the tubes also, in which case a specialist repair with undersize bushes must be made.
4 The fit within the lower fork legs is also important. If wear of the inner surface is evident, it may be necessary to fit lower bushes that have a slightly greater outside diameter.

7 Steering head bearings - examination and replacement

1 Before commencing to reassemble the forks, inspect the steering head races. The ball bearing tracks should be polished and free from indentations and cracks. If signs of wear are evident, the cones and cups must be replaced. They are a tight press fit and must be drifted out of position. A BSA Service Tool No. 61-3063 is available for extracting the cups that remain within the steering head assembly of the frame. It screws into the threaded centre of each cup and is driven out from the opposite end, bring the cup with it.
2 Ball bearings are cheap. If there is any reason to suspect the condition of the existing ball bearings, they should be replaced

without question. Note that each race is not completely full of ball bearings. Space should be left for the theoretical insertion of one extra ball, so that the race is not crowded, forcing the ball bearings to skid against one another.

3   Use thick grease to retain the ball bearings in position, whilst the head stem is being assembled and adjusted.

8   Front forks - reassembly

1   To reassemble the forks, follow the dismantling procedure in reverse. Take particular care when passing the sliding fork members through the oil seals, which should be fitted with the lip facing downwards. It is a wise precaution to wind a turn or so of medium twine around the undercut at the base of the thread of the plated collars, to act as an extra seal.

2   Tighten the steering head carefully, so that all play is eliminated without placing undue stress on the bearings. The adjustment is correct if all play is eliminated and the handlebars will swing to full lock of their own accord when given a light push on one end.

3   It is possible to place several tons pressure quite unwittingly on the steering head bearings, if they are over-tightened. The usual symptom of over-tight bearings is a tendency for the machine to roll at low speeds, even though the handlebars may appear to turn quite freely.

4   One problem that will arise during reassembly is the reluctance of the four main tubes to pass up into the fork top yoke. BSA Service Tool No. 61-3350 is used for this purpose; it threads into the top of each fork tube and can be used to pull the tube upward so that the tapered end engages with the fork yoke. If the tool is not available, a broom handle of the correct diameter can be used with the equal effect, if it is first screwed onto the end of the thread of each fork tube. Care should be taken in this instance, tp prevent particles of wood from falling into the fork tubes.

5   If, after assembly, it is found that the forks are incorrectly aligned or unduly stiff in action, loosen the front wheel spindle, the two caps at the top of the fork legs and the pinch bolts in both the top and bottom yokes. The forks should then be pumped up and down several times to realign them. Retighten all the nuts and bolts in the same order, finishing with the steering head pinch bolt.

6   This same procedure can be used if the forks are misaligned after an accident. Often the legs will twist within the fork yokes, giving the impression of more serious damage, even though no structural damage has occured.

7   Do not omit to add the correct amount of damping oil to each fork leg before replacing the fork leg caps. See specifications list for the amount and viscosity of oil to be added.

8   Although the Post 1970 slimline forks are different in appearance the same broad dismantling procedure applies.

9   Front forks - damping action

1   Each fork leg contains a predetermined quantity of oil to recommend viscosity, which is used as a damping medium to control the action of the compression springs within the forks when various road shocks are encountered. If the damping fluid is absent, there is no control over the rebound action of the fork springs and fork movement will be excessive, giving a very 'lively' ride. Damping restricts fork movement on the rebound and is progressive in action - the effect becomes more powerful as the rate of deflection increases.

2   In the BSA system, the oil is contained in the lower fork leg. When the forks are deflected, the space between the upper and lower fork bushes becomes greater and oil enters the inner fork tube, via the large diameter hole in the nut at the bottom of each fork tube, under force. Small diameter holes in the inner tube allow the oil to enter the space between the two fork bushes, but they have a restrictive effect, which slows down the movement

of the sliding tube. The rate of damping is governed by the size of these holes - the smaller the holes, the greater the damping effect. Before the sliding tube can reach its limit of travel, a tapered plug in the bottom of the lower fork leg enters the hole in the nut of the inner fork tube and slows down the rate of oil transfer until it is virtually cut off altogether. The remaining oil is incompressible and the fork leg is therefore prevented from 'bottoming' in a most effective manner. When the fork leg moves in the opposite direction (rebound) the space between the two fork bushes is reduced and the oil contained within this space must return through the holes drilled in the inner fork tube. As the fork movement increases still further, the uppermost bush will cover the holes completely, thereby leaving a quantity of oil in the space that cannot be compressed. The rebound action ceases as a result - it has be damped out.

3   The damping action can be varied only by changing the viscosity of the oil used as the damping medium; it is not practicable to vary the size of the holes in the inner fork tubes. In temperate climates an SAE 20 oil is used but if considered necessary, the viscosity rating can be increased without any harmful effects.

10 Frame assembly - examination and renovation

1   On both the earlier and present models a duplex cradle frame is utilised. Earlier models had a narrow frame tube from the top of the steering head to the rear frame; later models have a larger diameter 'spine' which offers increased steering head and swinging arm rigidity.

2   If the machine is stripped for an overhaul, this affords an excellent oppurtunity to inspect the frame for signs of cracks or other damage that may have occcured in service. Check the front down tubes at the point immediately below the steering head which is where a break is most likely to occur. Check the top tube of the frame for straightness - this is the tube most likely to bend in the event of an accident.

3   Frame repairs are best entrusted to a specialist in this field of repair work. It is often cheaper to buy a second hand frame from a vehicle breaker rather than get a bent one repaired.

11 Swinging arm rear suspension - examination and renovation

1   After an extended period of service, the bush and pivot pin of the swinging arm fork will wear, giving rise to lateral play that will affect the handling characteristics of the machine. The bushes are of the Silentbloc type, held in frame lugs close to the gearbox centre. They are a tight press fit and cannot be removed without risk of damage unless the correct equipment is available. This form of repair is best entrusted to a BSA repair specialist who will have the appropriate equipment. It is highly improbable that the average rider/owner will have access to this equipment or the skill with which to undertake the reconditioning work necessary.

2   The rear suspension units are removed by withdrawing the upper and lower bolts, nuts and washers.

12 Rear suspension units - examination

1   Only a limited amount of dismantling can be undertaken because the damper unit is sealed and cannot be dismantled. If the unit leaks oil, or if the damping action is lost, the unit must be replaced as a whole after removing the compression spring and shroud.

2   Before the spring and shroud can be removed, the unit must be detached from the machine and clamped in a vice. If pressure is applied to the top of the shroud, compressing the internal spring, the split collets can be removed and the spring and shroud released. Note the spring is colour-coded; the colour relates to the spring rating. Springs can be obtained in a variety of different ratings, to accommodate different loadings.

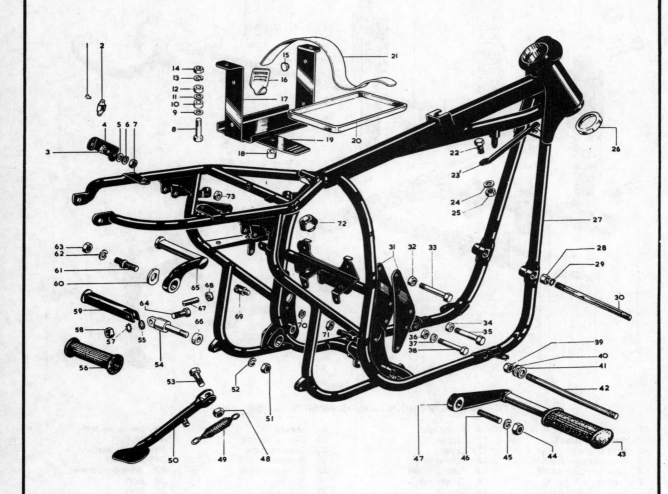

FIG. 4.4. FRAME ASSEMBLY (PRE 1971)

1 Rivet	19 Strap retainer	36 Nut	55 Washer
2 'Oddie' clip	20 Base rubber	37 Spring washer	56 Footrest rubber (pillion)
3 Bolt	21 Strap	38 Bolt (rear)	57 Washer
4 Bracket	22 Bolt	39 Nut	58 Nut
5 Washer	23 Stay	40 Washer	59 Footrest (pillion)
6 Spring washer	24 Washer	41 Washer	60 Lock plate
7 Nut	25 Nut	42 Stud	61 Stud
8 Bolt	26 Bearing cup	43 Footrest rubber	62 Spring washer
9 Washer	27 Frame	44 Nut	63 Nut
10 Rubber bush	28 Nut	45 Washer	64 Bolt
11 Washer	29 Washer	46 Stud	65 Footrest (left-hand)
12 Rubber bush	30 Stud	47 Footrest (right-hand)	66 Distance piece
13 Spring washer	31 Engine plate	48 Nut	67 Adjuster screw
14 Nut	32 Nut	49 Spring	68 Nut
15 Rubber plug	33 Bolt (lower)	50 Prop stand	69 Adjuster post
16 Hook	33 Bolt (top)	51 Nut	70 Spring washer
17 Battery carrier	34 Nut	52 Spring washer	71 Nut
18 Rubber bush	35 Bolt	53 Bolt	72 Rubber pad
		54 Anchor bolt	73 Distance piece (left-hand)

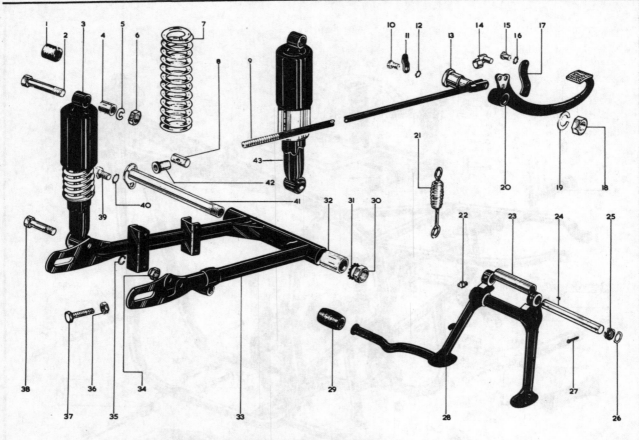

FIG. 4.5. SWINGING ARM COMPONENTS (PRE 1969)

1	Damper bush	9	Brake rod	21	Spring
2	Bolt (left-hand)	10	Screw	22	Grease nipple
2	Bolt (right-hand)	11	Lever	23	Spacing tube
3	Rear damper (solo)	12	Lockwasher	24	Spindle
3	Rear damper (solo)	13	Fulcrum pin	25	Spring washer
4	Spacer (left-hand)	14	Toggle pin	26	Washer
4	Spacer (right-hand)	15	Screw	27	Split pin
5	Spring washer	16	Washer	28	Centre stand
6	Nut	17	Bracket	29	Stand rubber
7	Spring (solo)	18	Nut	30	Nut
7	Spring (sidecar)	19	Spring washer	31	Washer
8	Swivel pin	20	Brake pedal	32	Silentbloc

33	Swinging arm
34	Nut
35	Spring washer
36	Locknut
37	Adjuster screw
38	Bolt
39	Bolt
40	Washer
41	Spindle
42	Adjuster nut
43	Rear damper (solo)
43	Rear damper (sidecar)

13 Rear suspension units - adjusting the setting

1   The Girling rear suspension units fitted to the BSA twin cylinder models have a three position cam adjuster built onto the lower portion of the leg to suit varying load conditions. The lowest position should suit the average rider, under normal road conditions. When a pillion passenger is carried, the second or middle position offers a better choice and for continuous high speed work of off-the-road riding, the highest position is recommended.

2   These adjustments can be effected without need to detach the units. A 'C' spanner in the tool kit is used to rotate the cam ring until the desired setting is obtained.

14 Centre stand - examination

1   All models, are provided with a centre stand attached to lugs on the bottom frame tubes. The stand provides a convenient means of parking the machine on level ground, or for raising one or the other of the wheels clear of the ground in the event of a puncture. The stand pivots on a long bolt that passes through the lugs and is secured by a nut and a washer. A return spring retracts the stand so that when the machine is pushed forward it will spring up and permit the machine to be wheeled, prior to riding.

2   The condition of the return spring and the return action should be checked frequently, also the security of the nut and bolt. If the stand drops whilst the machine is in motion, it may catch in some obstacle in the road and unseat the rider.

15 Prop stand - examination

1   A prop stand that pivots from a lug at the front end of the lower left hand frame tube provides an additional means of parking the machine. This too has a return spring, which should be strong enough to cause the stand to retract immediately the machine is raised into a vertical position. It is important that this spring is examined at regular intervals, also the nut and bolt that act as the pivot. A falling prop stand can have far more serious consequences if it should fall whilst the machine is on the move.

16 Footrests - examination and renovation

1 The footrests, which bolt to the frame lugs, are malleable and will bend if the machine is dropped. Before they can be straightened they must be detached from the frame and have the rubbers removed.

2 The footrests on early models are mounted on tapers and consequently can be positioned to suit the rider. Note, however, that although the right-hand footrest is secured by a nut and washer of normal thread, the left-hand footrest is secured by a nut with a left-hand thread. 'Oil-in-frame' frames (1971 to 1973) have footrests located by pegs and secured by nuts and bolts.

3 To straighten the footrests, clamp them in a vice and apply leverage from a long tube that slips over the end. The area in which the bend has occurred should be heated to a cherry red with a blow lamp, during the bending operation. Do not bend the footrests cold, otherwise there is risk of a sudden fracture.

17 Speedometer - removal and replacement

1 The earlier BSA twin cylinder models are fitted with a Smiths Chronometric speedometer, calibrated up to 120 mph. Later models are fitted with a magnetic speedometer head. An internal lamp is provided for illuminating the dial during the hours of darkness and the odometer has a trip setting, so that the lower mileage reading can be set to zero before a run is commenced.

2 The speedometer head has two studs, which permit it to be attached to a bracket on the fork top yoke. Removal of the retaining nut will permit the speedometer head to be withdrawn from the top of the mounting, after the drive cable has been detached.

3 Apart from defects in the drive or the drive cable itself, a speedometer that malfunctions is difficult to repair. Fit a replacement or alternatively entrust the repair to an instrument repair specialist, bearing in mind that the speedometer must function in a satisfactory manner to meet statutory requirements.

4 If the odometer readings continue to show an increase, without the speedometer indicating the road speed, it can be assumed the drive and drive cable are working correctly and that the speedometer head itself is at fault.

18 Speedometer cable - examination and renovation

1 It is advisable to detach the speedometer drive cable from time to time in order to check whether it is adequately lubricated, and whether the outer covering is compressed or damaged at any point along its run. A jerky or sluggish speedometer movement can often be attributed to a cable fault.

2 To grease the cable, withdraw the inner cable. After removing the old grease, clean with a petrol soaked rag and examine the cable for broken strands or other damage.

3 Regrease the cable with high melting point grease and ensure that there is no grease on the last six inches, at the end where the cable enters the speedometer head. If this precaution is not observed, grease will work into the speedometer head and immobilise the movement.

4 Inspection will show whether the speedometer drive cable has broken. If so, the inner cable can be removed and replaced with another whilst leaving the outer cable in place - provided the outer cable is not damaged or compressed at any point along its run. Measure the cable length exactly when purchasing a replacement, because this measurement is critical.

19 Tachometer - removal and replacement

1 The tachometer drive is normally taken off the front of the timing cover. The tachometer head may be of either the Chronometric or magnetic type, depending on the year of manufacture.

2 It is not possible to effect a satisfactory repair to a defective tachometer head, hence replacement is necessary if the existing head malfunctions. Make sure an exact replacement is obtained; some tachometer heads work at half-speed.

3 The tachometer head is illuminated internally so that the dial can be read during the hours of darkness.

20 Tachometer drive cable - examination and renovation

1 Although a little shorter in length, the tachometer drive cable is identical in construction to that used for the speedometer drive. The advice given in Section 18 of the Chapter applies also to the tachometer drive cable.

21 Removal of side covers

1 The glass fibre covers which on earlier models shroud the oil tank/tool kit compartment and battery are secured by two fasteners. To remove the covers give each fastener a half turn to release it or to lock it.

2 On post 1970 models with conical hubs the side panels have been redesigned.

3 To remove the right hand cover, first unscrew the master switch bezel. Take off the air filter cover, slacken the two attachment bolts from inside the air filter box and draw off the panel front edge first, and then upwards from its locating peg on the frame tube. When replacing do not omit the grommet from the peg.

4 The left side panel is removed in a similar manner, except for the absence of the master switch.

22 Dual seat removal

1 On models up to 1970 the dual seat is retained in position by two bolts with washers and distance pieces underneath the seat at each side of the mudguard. To remove, take out the bolts and washers noting the position of each distance piece and unhook the seat from the front. Replacement is the reversal of this procedure.

2 The redesigned models after 1970 have a hinged dual seat retained in position by a catch on the left side.

3 Open the seat and unbolt the front hinge. This will allow the seat to be removed from the rear hinge without further dismantling.

23 Tank badges and motifs

1 Different types of tank badge have been used since the inception of the BSA twin cylinder models, ranging from the BSA winged motif to circular or pear shaped plastics badges that either screw direct to the petrol tank or are retained by small hidden brackets.

2 The plastics badges are mounted over a thin rubber mat, to eliminate vibration chatter and to give some form of cushioning when the centre retaining screw is tightened. Certain colour schemes apply to certain models and if a badge is lost it is important to specify the colour scheme when endeavouring to obtain a replacement.

24 Steering head lock

1 All models are fitted with a steering head lock inserted into the fork top yoke. If the forks are turned to the extreme left, they can be locked in this position to prevent theft. The lock is of Yale manufacture.

2 Add an occasional few drops of thin machine oil to keep the lock in good working order. This should be added to the periphery of the moving drum and NOT the keyhole.

25 Steering damper

1 Mention has been made of the steering damper, a friction plate device that can be adjusted to vary the amount of effort

required to turn the handlebars. In may respects, the steering damper can be regarded as a legacy of the past when it was necessary to counteract the tendency of some machines to develop a speed wobble at high speeds. Today the steering damper comes into its own mainly when a sidecar is attached, since it will prevent the handlebars from oscillating at very low speeds, when it is applied.

2 Under normal riding conditions, the steering damper can be slackened off. It is sometimes advantageous to have it biting just a trifle at high speeds, to ease the strain on the arms.

3 To remove the steering damper, detach the split pin at the extreme end of the damper rod and/or unscrew the knob at the top of the steering head until it can be drawn away with the rod attached. When the fixed plate is detached from below the bottom yoke of the forks, the friction plates can be removed for inspection. They seldom require attention.

26 Cleaning - general

1 After removing all surface dirt with a rag or sponge that is washed frequently in clean water, the application of car polish or wax will restore a good finish to the cycle parts of the machine after they have dried thoroughly. The plated parts should require only a wipe with a damp rag, although it is permissible to use a chrome cleaner if the plated surfaces are badly tarnished.

2 Oil and grease, particularly when they are caked on, are best removed with a proprietary cleanser such as 'Gunk' or 'Jizer'. A few minutes should be allowed for the cleanser to penetrate the film of oil and grease before the parts concerned are hosed down. Take care to protect the magneto, carburettor(s) and electrical parts from the water, which may otherwise cause them to malfunction.

3 Polished aluminium alloy surfaces can be restored by the application of Solvol 'Autosol' or some similar polishing compound, and the use of a clean duster to give the final polish.

4 If possible, the machine should be wiped over immediately after it has been used in the wet, so that it is not garaged under damp conditions that will promote rusting. Make sure to wipe the chain and if necessary re-oil it, to prevent water from entering the rollers and causing harshness with an accompanying high rate of wear. Remember there is little chance of water entering the control cables if they are lubricated regularly, as recommended in the Routine Maintenance Section.

27 Fault diagnosis

Symptom	Cause	Remedy
Machine is unduly sensitive to road conditions	Forks and/or rear suspension units have defective damping	Check oil level in forks. Replace rear suspension units.
Machine tends to roll at low speeds	Steering head bearings overtight or damaged	Slacken bearing adjustment. If no improvement, dismantle and inspect bearings.
Machine tends to wander, steering is imprecise	Worn swinging arm bearings or sliders in plunger sprung models	Check and if necessary renew bearings.
Fork action stiff	Fork legs have twisted in yokes or have been drawn together at lower ends	Slacken off spindle nut clamps, pinch bolts in fork yokes and fork top nuts. Pump forks several times before retightening from bottom. Is distance piece missing from fork spindle?
Forks judder when front brake is applied	Worn fork bushes Steering head bearings too slack	Strip forks and replace bushes. Readjust, to take up play.
Wheels out of alignment	Frame distorted as result of accident damage	Check frame alignment after stripping out. If bent, specialist repair is necessary.

Chapter 5 Wheels, brakes and tyres

Contents

Specifications

Wheels:

Rim sizes - front		WM2 x 19 or WM3 x 19
- rear		WM2 x 18 or WM3 x 18
Tyre sizes - front		3.25 x 19, 3.50 x 19, 4.00 x 19 *
- rear		4.00 x 18

Brakes:

Type - front	Single leading shoe - drum
	twin leading shoe - drum
- rear	Single leading shoe - drum
Diameter - front	8 in. single leading shoe
	190 mm (Spitfire)
	8 in. twin leading shoes 1968 on
- rear	7 in. drum

Final drive chain:

Size	5/8 in. x 3/8 in.
Number of pitches	106, 110
Rear wheel sprocket - No. of teeth	47

Tyre pressure — tyres cold*:

Front	21 p.s.i.
Rear	22 p.s.i.

Note that the pressures given are the basic setting for low speed with a rider of 154 lb. For high speed increase both by 5 psi — for heavier loads add 1 psi (front) or 2 psi (rear) for every extra 28 lb load

1 General description

The unit construction twin cylinder models are fitted with a 19" diameter front wheel and an 18" diameter rear wheel. All models have a 3.25" section tyre fitted to the front wheel; the rear wheel has either a 3.50" section tyre, or in the case of the post 1970 models, a tyre of 4.00" section. The American export models are fitted as standard with the larger section rear tyre.

Until 1971 it was customary to fit either a 190 mm internal expanding brake of the single leading shoe type or twin leading shoe brake of 8" diameter to the front wheel and a 7" diameter internal expanding brake of conventional design to the rear wheel. Thereafter an 8" diameter front brake of the twin leading shoe type, with its adjustment modelled on car practice but cable operated, was substituted for the original design, whilst the rear brake remained virtually unchanged. These models are identified by the conical shape of the hubs.

The rear wheel is of the quickly detachable variety which means the wheel can be removed, if desired, leaving the final drive chain and sprocket in position. This arrangement does not apply to later models fitted with concical hubs.

2 Front wheel - examination and removal

1 Place the machine on the centre stand so that the front wheel is raised clear of the ground. Spin the wheel and check for rim alignment. Small irregularities can be corrected by tightening the spokes in the affected area, although a certain amount of skill is necessary if over-correction is to be avoided. Any 'flats' in the wheel rim should be evident at the same time. These are more difficult to remove with any success and in most cases the wheel will have to be rebuilt on a new rim. Apart from the effect on stability, there is a greater risk of damage to the tyre bead and walls if the machine is run with a deformed wheel, especially at high speeds.

2 Check for loose or broken spokes. Tapping the spokes is the best guide to the correctness of tension. A loose spoke will produce a quite different note and should be tightened by turning the nipple in an anti-clockwise direction. Always check for run-out by spinning the wheel again.

3 If several spokes require retensioning or there is one that is particularly loose, it is advisable to remove the tyre and tube so that the end of each spoke that projects through the nipple after retensioning can be ground off. If this precaution is not taken, the portion of the spoke that projects may chafe the inner tube and cause a puncture.

3 Front drum brake assembly - examination, renovation and reassembly

1 The front brake assembly complete with brake plate can be withdrawn from the front wheel by following the procedure in Chapter 6, section 2, paragraphs 1 to 3. In the case of the post 1970 models, fitted with conical hubs, a slightly different procedure is necessary when detaching the brake cable from the rearmost brake operating arm. It is also necessary to slacken the torque lug nut on the inner portion of the right hand fork leg before the wheel can be freed.

2 An anchor plate nut retains the brake plate on the front wheel spindle. When this nut is removed, the brake plate can be drawn away, complete with the brake shoe assembly.

3 Examine the condition of the brake linings. If they are wearing thin or unevenly, the brake shoes should be relined or renewed.

4 To remove the brake shoes from the brake plate, pull them apart whilst lifting them upward, in the form of a V. When they clear of the brake plate, the return springs can be removed and the shoes separated. Do not lose the abutment pads fitted to the leading edge of each shoe.

5 The brake linings are rivetted the brake shoes and it is easy to remove the old linings by cutting away the soft metal rivets. If the correct BSA replacements are purchased, the new linings will be supplied ready drilled with the correct complement of rivets. Keep the lining tight against the shoe throughout the rivetting operation and make sure the rivets are countersunk well below the lining surface. If workshop facilities and experience suggest it would be preferable to obtain replacement shoes ready lined, costs can be reduced by making use of the BSA service exhange scheme, available through BSA agents.

6 Before replacing the brake shoes, check that the brake operating cam(s) are working smoothly and not binding in the pivot(s). The cam(s) can be removed for cleaning and greasing by unscrewing the nut on each brake operating arm and drawing the arm off, after its position relative to the cam spindle has been marked so that it is replaced in exactly the same position. The spindle and cam can then be pressed out of the housing in the back of the brake plate.

7 Check the inner surface of the brake drum on which the brake shoes bear. The surface should be smooth and free from score marks or indentations, otherwise reduced braking efficiency is inevitable. Remove all traces of brake lining dust and wipe both the brake drum surface and the brake shoes with a clean rag soaked in petrol, to remove any traces of grease. Check that the brake shoes have chamfered ends to prevent pick-up or grab. Check that the brake shoe return springs are in good order and have not weakened.

8 To reassemble the brake shoes on the brake plate, fit the return springs first and force the shoes apart, holding them in a V formation. If they are now located with the operating cams they can usually be snapped into position by pressing downward. Do not use excessive force or the shoes may distort permanently. Make sure the abutment pads are not omitted.

9 A different type of brake unit is fitted to the post 1970 models which have wheels with conical hubs. Although the operating principle is the same, car-type brake shoe expanders are fitted. A micram adjuster is fitted in place of the abutment pads used previously, providing a means of compensating for brake lining wear without having to reduce the angle of the brake operating arms. The brake unit can be dismantled and reassembled by using the procedure already described.

3.1 Brake plate will lift away when front wheel is detached from forks

3.4 Pull and lift brake shoe upwards to separate from brake plate

4 Front wheel bearings - removal, examination and replacement

1 When the brake plate of machines fitted with a drum brake is removed, the bearing retainer within the brake drum will be exposed. Pre 1971 models had bearing retainers with a normal right hand thread and these were locked in position by split pins. Post 1971 models had a left hand threaded bearing retainer on the right hand side (brake shoe side) and a circlip on the left hand side of the hub.

2 Remove the split pins if fitted and unscrew the bearing retainers the appropriate way either with a service tool (No. 61-3694) or by tapping round with a screwdriver or punch.

3 Replace the front wheel spindle through the left hand front wheel bearing and drive out the brake side bearing together with the bush by striking the head of the spindle with a hide mallet.

4 The left hand bearing can be driven out in a similar way with the spindle and bush driven in from the right hand side. Both bearings are the same size and therefore interchangeable.

5 After removing both bearings wash them in paraffin and let them dry. Examine the bearings for excessive play and roughness in the bearing tracks; If necessary renew them. Any bearing which is replaced should be carefully packed with new high melting point grease.

6 Replacement bearings are simply fitted in the reverse manner, but pressure must only be applied to the outer ring of the bearing.

7 All models have rubber grease retainers and these should be checked and replaced only if in good condition. Earlier models had one grease retainer behind each bearing retaining screw whereas post 1971 models have grease retainers on both sides of each wheel bearing.

8 Note that the retainer with the large hole is fitted on the brake side and remember to replace the bush when the right hand wheel bearing is replaced.

4.2 Front wheel bearing retainer has a right hand thread (pre 1971)

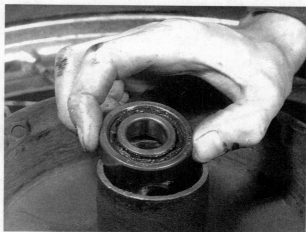

4.3 Bearing will drive out of hub

5 Front wheel replacement

1 For earlier models with a quick release hub lift the wheel between the fork legs and position the bush in the right hand fork leg.

2 Screw the spindle in anti-clockwise (left hand thread) until it is nearly tight and position the brake plate and replace the brake anchor strap with its appropriate nuts, bolts and washers.

3 Tighten the wheel spindle fully and lock the wheel spindle by tightening the pinch bolt on the left hand fork end.

4 Replace the front brake cable and adjust as necessary using the adjuster on the brake cover plate.

5 On later models with a twin leading shoe front brake and/or conical front hub lift the wheel between the forks and locate the peg on the fork leg to the slot on the brake plate, at the same time locating the spindle in the fork bottoms. Replace the fork caps and the two clamp bolts on each fork leg bottom.

6 Tighten all four clamp bolts fully and replace the front brake cable.

7 Models with a conical front hub have a threaded peg on the brake plate which locates in a slotted ear on the front fork outer member. Locating the front wheel in the forks is done in a similar manner to the twin leading shoe hub expect that instead of locating a peg on the fork leg in the brake plate, the threaded peg on the brake plate is slotted in the ear on the front fork leg and secured with a nut and spring washer.

8 Replace the brake cable and check that the brake functions properly, especially if the adjustment has been altered or the brake operating arms have been removed and replaced during the dismantling operation.

4.3a ... followed by hollow spacer

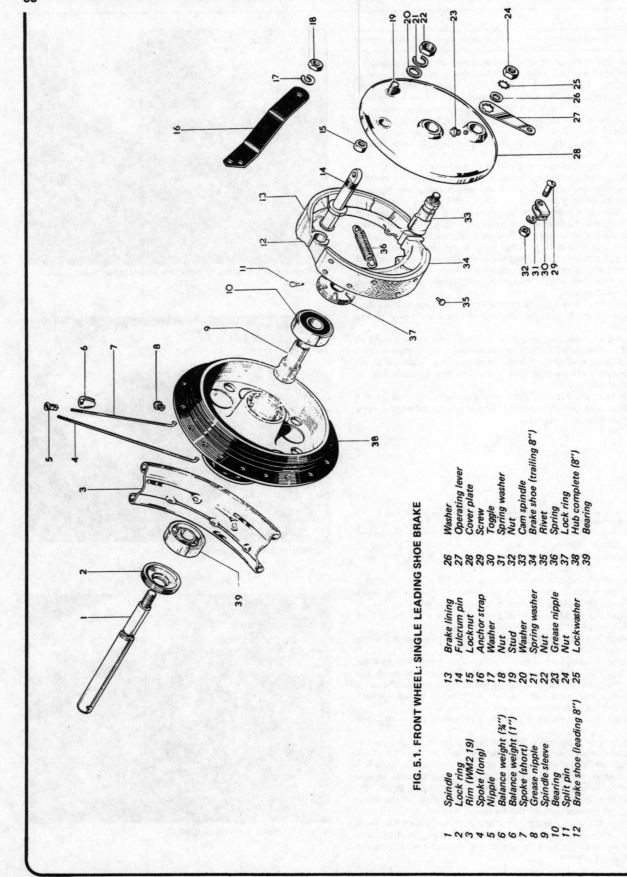

FIG. 5.1. FRONT WHEEL: SINGLE LEADING SHOE BRAKE

1	Spindle	13	Brake lining	26	Washer
2	Lock ring	14	Fulcrum pin	27	Operating lever
3	Rim (WM2 19)	15	Locknut	28	Cover plate
4	Spoke (long)	16	Anchor strap	29	Screw
5	Nipple	17	Washer	30	Toggle
6	Balance weight (¾")	18	Nut	31	Spring washer
6	Balance weight (1")	19	Stud	32	Nut
7	Spoke (short)	20	Washer	33	Cam spindle
8	Grease nipple	21	Spring washer	34	Brake shoe (trailing 8")
9	Spindle sleeve	22	Nut	35	Rivet
10	Bearing	23	Grease nipple	36	Spring
11	Split pin	24	Nut	37	Lock ring
12	Brake shoe (leading 8")	25	Lockwasher	38	Hub complete (8")
				39	Bearing

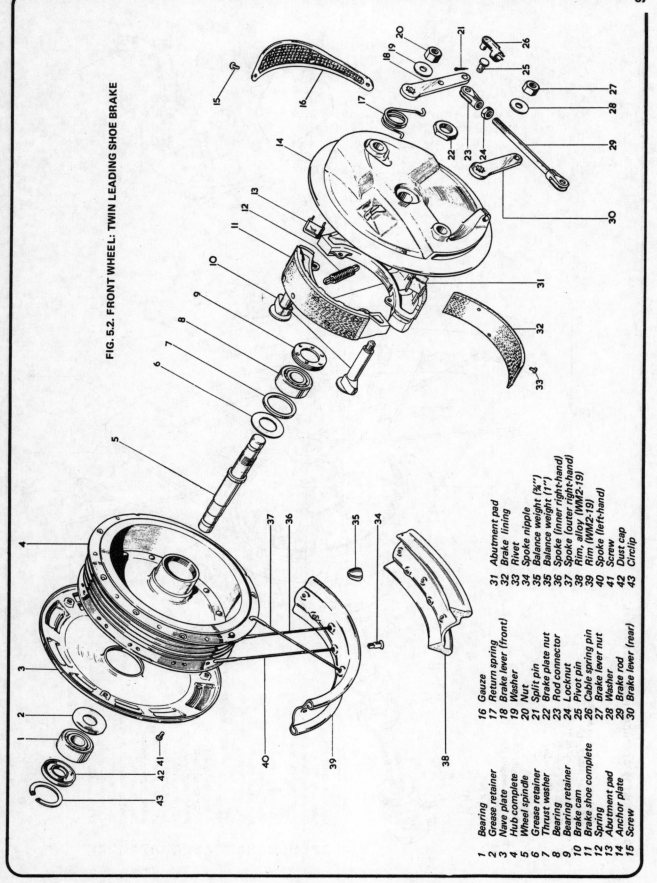

FIG. 5.2. FRONT WHEEL: TWIN LEADING SHOE BRAKE

1 Bearing
2 Grease retainer
3 Nave plate
4 Hub complete
5 Wheel spindle
6 Grease retainer
7 Thrust washer
8 Bearing
9 Bearing retainer
10 Brake cam
11 Brake shoe complete
12 Spring
13 Abutment pad
14 Anchor plate
15 Screw

16 Gauze
17 Return spring
18 Brake lever (front)
19 Washer
20 Nut
21 Split pin
22 Brake plate nut
23 Rod connector
24 Locknut
25 Pivot pin
26 Cable spring pin
27 Brake lever nut
28 Washer
29 Brake rod
30 Brake lever (rear)

31 Abutment pad
32 Brake lining
33 Rivet
34 Spoke nipple
35 Balance weight (¾")
35 Balance weight (1")
36 Spoke (inner right-hand)
37 Spoke (outer right-hand)
38 Rim, alloy (WM2-19)
39 Rim (WM2-19)
40 Spoke (left-hand)
41 Screw
42 Dust cap
43 Circlip

FIG. 5.3. POST 1970 CONICAL HUB FRONT WHEEL

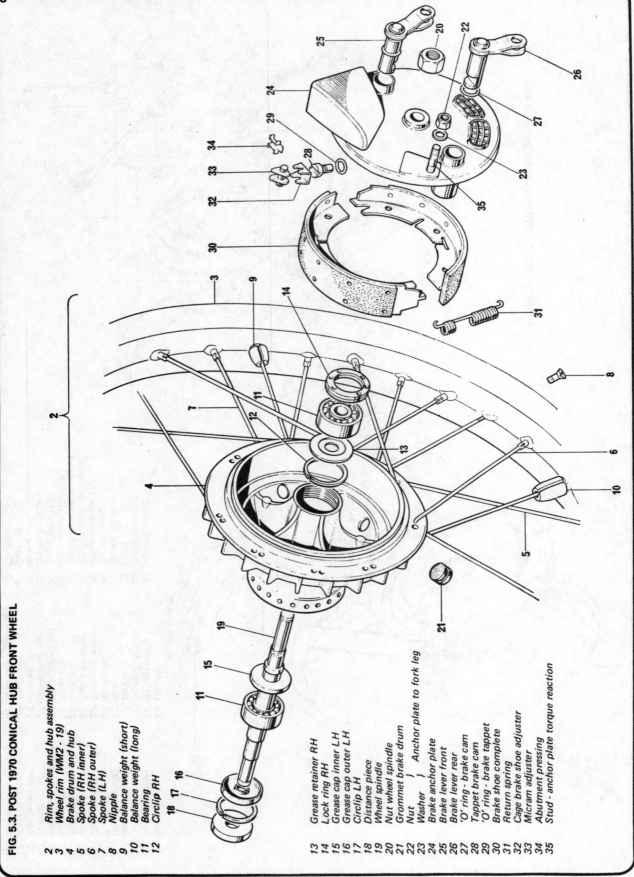

2   Rim, spokes and hub assembly
3   Wheel rim (WM2 - 19)
4   Brake drum and hub
5   Spoke (RH inner)
6   Spoke (RH outer)
7   Spoke (LH)
8   Nipple
9   Balance weight (short)
10 Balance weight (long)
11 Bearing
12 Circlip RH

13 Grease retainer RH
14 Lock ring RH
15 Grease cap inner LH
16 Grease cap outer LH
17 Circlip LH
18 Distance piece
19 Wheel spindle
20 Nut wheel spindle
21 Grommet brake drum
22 Nut   ) Anchor plate to fork leg
23 Washer)
24 Brake anchor plate
25 Brake lever front
26 Brake lever rear
27 'O' ring - brake cam
28 Tappet brake cam
29 'O' ring - brake tappet
30 Brake shoe complete
31 Return spring
32 Cage brake shoe adjuster
33 Micram adjuster
34 Abutment pressing
35 Stud - anchor plate torque reaction

6 Rear wheel - removal and examination

1 Place the machine on the centre stand and before removing the wheel, check for rim alignment, loose or broken spokes and other defects by following the procedure applying to the front wheel in Section 2.

2 Two types of rear wheel have been fitted, the standard or the quickly detachable type. The latter has the advantage of simplified removal, leaving the final drive chain and sprocket in position. The option is no longer available with the latest conical hub wheel which is not made in the quickly detachable version.

3 If the wheel is of the standard type, commence by disconnecting the final drive chain at the detachable spring link. The task is made easier if the link is first positioned so that it is on the rear wheel sprocket. Unwind the chain off the rear sprocket and lay it on a clean surface.

4 Take off the brake rod adjuster and pull the brake rod clear of the brake operating arm. Disconnect the torque stay by removing the nut where the stay joins the rear brake plate. If the stop lamp stays in the 'on' position disconnect the snap connector in the lead.

5 Slacken both rear wheel spindle nuts and raise the rear chain guard by slackening the bottom nut of the left hand suspension unit. Remove the speedometer cable from the gearbox on the right hand side of the rear hub (if fitted) by unscrewing the gland nut and withdrawing the cable. Withdraw the wheel rearward until it drops from the frame ends complete with chain adjuster. It may be necessary to tilt the machine or raise it higher so that there is sufficient clearance for the wheel to be taken away from the machine.

6 A different procedure is employed in the case of machines fitted with the quickly detachable rear wheel. It is only necessary to unscrew the gland nut from the speedometer drive and withdraw the cable, then unscrew and remove the wheel spindle from the right hand side of the machine. If the shouldered distance piece between the frame end and the hub is removed, the wheel can be pulled sideways to disengage it from the brake drum centre (splined fitting) before it is lifted away. Note there is a rubber seal over the splines which is compressed when the wheel is in position. This acts as a grit seal and must be maintained in a good condition.

7 Rear wheel bearings - removal, examination and replacement

1 The procedure for removing the rear wheel bearings is similar to that of the front wheel.

2 With the rear wheel removed from the frame access to the right hand rear wheel bearing is gained by removing the speedometer drive gearbox. Removal of the speedometer gearbox reveals the bearing retainer ring which must be unscrewed. Models with a quick release rear hub have a bearing retainer ring with the normal right hand thread; Models would appear to vary, some having a bearing retainer with a left hand thread, 'L' being stamped on the retainer. Always check before unscrewing, therefore unscrew anticlockwise. Late models with the conical rear hub have a left hand thread on the retaining ring; therefore unscrew turning clockwise.

3 Remove the bearing retainer ring on the left hand side (brake drum) side. Quick release hub models also have a left hand thread. To gain access to the bearing retainer ring on late models the brake plate complete with shoes will first have to be removed.

4 On earlier models remove the bearings by first drifting out the hollow spindle from the left hand side, releasing the right hand bearing, distance piece, inner collar, pen steel washer and the dust covers from the right hand side.

5 The left hand bearing together with its thrust washer can now be driven out from the right hand side leaving the rubber seal in the hub which need not be disturbed.

6 On models fitted with the conical rear hub remove the bearings by first drifting out the hollow centre sleeve from the left side taking with it the right hand bearing. Remove the bearing from the sleeve and reinsert the sleeve into the left side bearing releasing its abutment ring at the same time.

7 Wash the bearings in paraffin and blow out with an air line if possible. Check for roughness in the bearing tracks and excessive play. If necessary renew.

8 Pack the bearings with new high melting point grease and replace them by reversing the dismantling procedure.

8 Rear brake - removal and examination

1 If the rear wheel is of the standard typre or if the rear wheel has been removed together with the brake drum and sprocket the rear brake assembly is accessible when the brake plate is lifted away from the drum.

2 If the quickly detachable wheel has been removed in the recommended way it will be necessary to detach the brake drum and sprocket assembly from the frame. This is accomplished by removing the final drive chain by detaching the spring link, unscrewing and removing the nut from the brake plate torque stay so that the latter can be pulled away and unscrewing and removing the large nut around the bearing sleeve that supports the brake and sprocket assembly. Note that the brake drum has a bearing in its centre which should be knocked out, cleaned and examined before replacement.

3 The rear brake assembly is similar to that of the front wheel drum brake apart from the fact that it is of the single leading shoe type and therefore has only one operating arm. Use an identical procedure for examining and renovating the brake assembly to that described in Section 3, paragraphs 3 to 7.

9 Rear brake replacement

1 Reverse the dismantling procedure when replacing the front brake to give the correct sequence of operations. In the case of the quickly detachable rear wheel, the brake drum and sprocket should be fitted to the frame first to aid assembly.

10 Rear wheel and gear box final drive sprockets - examination

1 Before replacing the rear wheel it is advisable to examine the rear wheel sprocket. A badly worn sprocket will greatly accelerate the wear of the final drive chain and in an extreme case, will even permit the chain to ride over the teeth when the initial drive is taken up. Wear will be self evident in the form of shallow or hooked teeth, indicating the need for early renewal.

2 In the case of the standard wheel the sprocket is secured by ten bolts which are threaded into the brake drum. On the latest models the sprocket is secured by five self locking nuts to the brake drum. When a quickly detachable wheel is fitted the sprocket is an integral part of the brake drum in which case the complete unit must be renewed.

3 The gearbox sprocket should also be inspected closely at the same time because it is considered bad practise to renew the one sprocket alone. A certain amount of dismantling work is necessary before the gearbox sprocket can be removed. Chapter 1.8 paragraphs 1 - 10 provides the relevant information.

11 Final drive chain - examamination and lubrication

1 The final drive chain is not fully enclosed. The only lubrication provided takes the form of a drip feed which relies on oil being collected in a small well at the back of the primary chaincase.

2 Chain adjustment is correct when there is approximately ¾" play in the middle of the chain run, measured at either the top or the bottom. Always check at the tightest spot of the chain run with the rider seated normally.

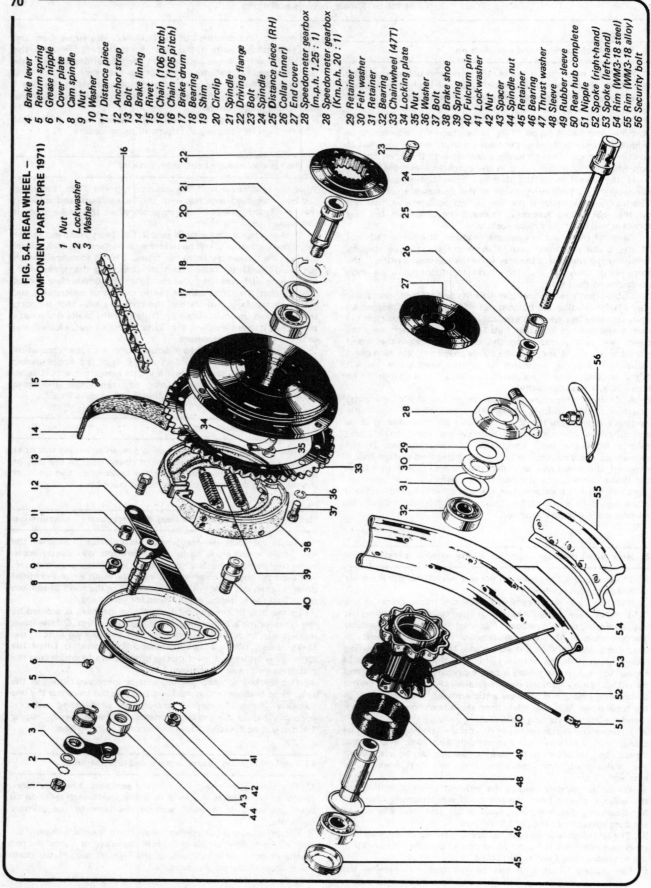

FIG. 5.4. REAR WHEEL — COMPONENT PARTS (PRE 1971)

1 Nut
2 Lockwasher
3 Washer
4 Brake lever
5 Return spring
6 Grease nipple
7 Cover plate
8 Cam spindle
9 Nut
10 Washer
11 Distance piece
12 Anchor strap
13 Bolt
14 Brake lining
15 Rivet
16 Chain (106 pitch)
16 Chain (105 pitch)
17 Brake drum
18 Bearing
19 Shim
20 Circlip
21 Spindle
22 Driving flange
23 Bolt
24 Spindle
25 Distance piece (RH)
26 Collar (inner)
27 End cover
28 Speedometer gearbox (m.p.h. 1.25 : 1)
28 Speedometer gearbox (Km.p.h. 20 : 1)
29 Retainer
30 Felt washer
31 Retainer
32 Bearing
33 Chainwheel (47T)
34 Locking plate
35 Nut
36 Washer
37 Bolt
38 Brake shoe
39 Spring
40 Fulcrum pin
41 Lockwasher
42 Nut
43 Spacer
44 Spindle nut
45 Retainer
46 Bearing
47 Thrust washer
48 Sleeve
49 Rubber sleeve
50 Rear hub complete
51 Nipple
52 Spoke (right-hand)
53 Spoke (left-hand)
54 Rim (WM3-18 steel)
55 Rim (WM3-18 alloy)
56 Security bolt

3 If the chain is too slack, adjust by slackening the wheel spindle and/or wheel nuts and the nut of the torque arm stay, then drawing the wheel rearward by the chain adjusters at the end of each swinging arm fork. It is important to ensure that each adjuster is turned the same amount so that the rear wheel is kept in alignment. When the correct adjusting point has been reached, push the wheel forward and tighten the wheel nuts and/or spindle, not forgetting the torque arm nut. Re-check the chain tensioner and wheel alignment before the final tightening.

4 To check whether the chain needs renewing, lay it lengthwise in a straight line and compress it end-wise in the opposite direction and measure the amount of stretch. If it exceeds ¼ inch per foot renewal is necessary. Never use an old or worn chain when new sprockets are fitted; it is advisable to renew the chain at the same time so that all parts run together.

5 Every 2000 miles remove the chain and clean it thoroughly in a bath of paraffin before immersing it in a special chain lubricant such as Linklyfe or Chainguard. These latter types of lubricant are applied in the molten state (the chain is immersed) and therefore achieve much better penetration of the chain links and rollers. Furthermore, the lubricant is less likely to be thrown off when the chain is in motion.

6 When replacing the chain, make sure the spring link is positioned correctly, with the closed end facing the direction of travel. Replacement is made easier if the end of the chain are pressed into the teeth of the rear wheel sprocket whilst the connecting link is inserted or a simple 'chain joiner' is used.

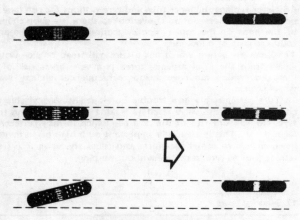

Fig. 5.5. Checking wheel alignment

12 Front brake adjustment

1 Brake adjustment is effected by the cable adjuster built into the front brake lever, which should be screwed outward to take up slack which develops in the operating cable as the brake shoes wear. Although adjustment is a matter of personal setting there should never be sufficient slack in the cable to permit the lever to touch the handlebars before the brake is applied fully.

2 Eventually braking action will be lost because cable adjustment has resulted in a poor angle between the brake operating arms and the direction of pull, causing loss of leverage. Provided the brake shoes are not badly worn, this can be corrected by slackening the adjuster fully and placing the machine on the centre stand. Remove the rubber plug from the front brake plate and insert a screwdriver so that each micram adjuster can be adjusted in turn. Start by turning the adjuster as far as it will go, so that the brake shoes are in contact with the brake drum. Back off two flats and check the wheel is free to revolve. Turn the wheel half a revolution and repeat with the second adjuster. A further small adjustment may be necessary with the handlebar

lever, to position the point of operation to the riders liking. Do not omit to replace both rubber plugs in the brake plate and to recheck the brake before the machine is used on the road. This additional method of adjustment applies to machines fitted with conical hub only.

3 On models fitted with the twin leading shoe brake (post 1968) a slightly different form of brake adjustment is needed to gain maximum efficiency from the brake after fitting new shoes or general renovation. The shoes have to be balanced in order to equalize their braking effort. With the front wheel in position with its brake cable attached disconnect the tie-rod between the brake levers of th brake plate mechanism by taking out the pivot pins at each end slacken the locknut on the tie rod. Apply the front brake and keep it applied by means of a strong rubber band or tape wrapped round the handlebar lever. Turn the short (rearmost) lever in its normal direction of travel by means of a spanner applied to the spindle nut, until the brake shoe is firmly in contact with the brake drum. Then adjust the length of the tie-rod until the pivot pins can be inserted at both ends of the rod. Fit new split pins to the pivot pins and retighten the locknut on the tie-rod.

13 Rear brake adjustment

1 Rear brake adjustment is effected solely by the screwed adjuster at the extreme end of the brake operating rod. It should be screwed inward to decrease the amount of play at the brake pedal. Always check after making adjustments to ensure that the brake shoes are not binding.

2 Brake adjustment will be necessary when slack in the chain is taken up. Because this involves moving the rear wheel backward in the frame the rear brake adjuster may have to be slackened off a little.

3 After the rear brake has been adjusted check the stop light action. It may be necessary to re-adjust the point at which the bulb lights by repositioning the clamp around the brake operating rod connected to the operating spring.

14 Front wheel balancing

1 It is customary on all high performance machines to balance the front wheel complete with tyre and tube. The out of balance forces which exist are then eliminated and the handling of the machine improved. A wheel, which is badly out of balance produces throughout the steering, a most unpleasant hammering effect at high speeds.

2 One ounce and half ounce balance weights are available which can be slipped over the spokes and engaged with the square section of the spoke nipples. The balance weights should be added oppostite to this point. Add or subtract balance weights until the wheel will rest in any position after it has been spun.

3 If balance weights are nto available, wire solder wrapped around the spokes, close to the nipples, is an excellent substitute.

5 There is no necessity to balance the rear wheel for normal road use.

15 Speedometer drive gearbox - general

1 Models have speedometer drive taken from the speedometer drive gearbox fitted to the right hand side of the rear wheel. The drive is transmitted from the hub by means of a slotted locking which threads into the hub and performs the dual function of retaining the right hand wheel bearing.

2 Provided this gearbox is greased at regular intervals it is unlikely to require attention during the normal life of the machine.

3 Speedometer drive gearboxes are not necessarily interchangeable even though they may look similar. If a replacement

has to be made, it is advisable to check the specification. The drive ratio is related to the size of the rear wheel and the section of tyre fitted, two variables which will have a marked effect on the accuracy of the speedometer reading.

16 Tyres - removal and replacement

1 At some time or other the need will arise to remove and replace the tyres, either as the result of a puncture or because a renewal is required to offset wear. To the inexperienced, tyre changing represents a formidable task yet if a few simple rules are observed and the technique learned the whole operation is surprisingly simple.

2 To remove the tyre from either wheel, first detach the wheel from the machine by following the procedure given in this Chapter whether the front or the rear wheel is involved. Deflate the tyre by removing the valve insert and when it is fully deflated, push the bead of the tyre away from the wheel rim on both sides so that the bead enters the centre well of the rim. Remove the locking cap and push the tyre valve into the tyre.

3 Insert a tyre lever close to the valve and lever the edge of the tyre over the outside of the wheel rim. Very little force should be necessary; if resistance is encountered it is probably due to the fact that the tyre beads have not entered the well of the wheel rim all the way round the tyre.

4 Once the tyre has been edged over the wheel rim, it is easy to work around the wheel rim so that the tyre is completely free on one side. At this stage, the inner tube can be removed.

5 Working from the other side of the wheel, ease the other edge of the tyre over the outside of the wheel rim furthest away. Continue to work around the rim until the tyre is free completely from the rim.

6 If a puncture has necessitated the removal of the tyre, re-inflate the inner tube and immerse it in a bowl of water to trace the source of the leak. Mark its position and deflate the tube. Dry the tube and clean the area around the puncture with a petrol soaked rag. When the surface has dried, apply rubber solution and allow this to dry before removing the backing from a patch and applying the patch to the surface.

7 It is best to use a patch of the self-vulcanising type, which will form a very permanent repair. Note that it may be necessary to remove a protective covering from the top surface of the patch, after it has sealed in position. Inner tubes made from synthetic rubber may require a special type of patch and adhesive if a satisfactory bond is to be achieved.

8 Before replacing the tyre, check the inside to make sure that the agent which caused the puncture is not trapped. Check the outside of the tyre, particularly the tread area, to make sure nothing is trapped that may cause a further puncture.

9 If the inner tube has been patched on a number of past occasions or if there is a tear or large hole, it is preferable to discard it and fit a new tube. Sudden deflation may cause an accident, particularly if it occurs with the front wheel.

10 To replace the tyre, inflate the inner tube just sufficiently for it to assume a circular shape. Then push it into the tyre so that it

is enclosed completely. Lay the tyre on the wheel at an angle and insert the valve through the rim tape and the hole in the wheel rim. Attach the locking cap on the first few threads, sufficient to hold the valve captive in its correct location.

11 Starting at the point furthest from the valve, push the tyre bead over the edge of the wheel rim until it is located in the central well. Continue to work around the tyre in this fashion until the whole of one side of the tyre is on the rim. It may be necessary to use a tyre lever during the final stages.

12 Make sure there is no pull on the tyre valve and again commencing with the area furthest from the valve, ease the other bead of the tyre over the edge of the rim. Finish with the area close to the valve, pushing the valve up into the tyre until the locking cap touches the rim. This will ensure the inner tube is not trapped when the last section of the bead is edged over the rim with a tyre lever.

13 Check that the inner tube is not trapped at any point. Re-inflate the inner tube, and check that the tyre is seating correctly around the wall of the tyre on both sides, which should be equidistant from the wheel rim at all points. If the tyre is unevenly located on the rim, try bouncing the wheel when the tyre is at the recommended pressure. It is probable that one of the beads has not pulled clear of the centre well.

14 Always run the tyres at the recommended pressures and never under or over-inflate. See Specifications for recommended pressures.

15 Tyre replacement is aided by dusting the side walls, particularly in the vicinity of the beads, with a liberal coating of French chalk. Washing-up liquid can also be used to good effect, but this has the disadvantage of causing the inner surfaces of the wheel rim to rust.

16 Never replace the inner tube and tyre without the rim tape in position. If this precaution is overlooked there is a good chance of the ends of the spoke nipples chafing the inner tube and causing a crop of punctures.

17 Never fit a tyre which has a damaged tread or side walls. Apart from the legal aspects, there is a very great risk of a blow-out, which can have serious consequences on any two-wheel vehicle.

18 Tyre valves rarely give trouble but it is always advisable to check whether the valve itself is leaking before removing the tyre. Do not forget to fit the dust cap which forms an effective second seal. This is especially important on a high performance machine, where centrifugal force can cause the valve insert to retract and the tyre to deflate without warning.

17 Security bolts

1 It is often considered necessary to fit a security bolt to the rear wheel of a high performance model becuase the initial take up of drive may cause the tyre to creep around the wheel rim and tear the valve from the inner tube. The security bolt retains the bead of the tyre to the wheel rim and prevents this occurence.

18 Fault diagnosis

Symptom	Cause	Remedy
Handlebars oscillate at low speeds	Buckle or flat in wheel rim, most probably front wheel	Check rim alignment by spinning wheel. Correct by retensioning spokes or rebuilding on new rim.
	Tyre not straight on rim	Check tyre alignment.
Machine lacks power and accelerates poorly	Brakes binding	Warm brake drum provides best evidence. Re-adjust brakes.
Brakes grab when applied gently	Ends of brake shoes not chamfered	Chamfer with file.
	Elliptical brake drum	Lightly skim in lathe (specialist attention required)
Brake pull-off sluggish	Brake cam binding in housing	Free and grease.
	Weak brake shoe springs	Renew if springs have not become displaced.
Harsh transmission	Worn or badly adjusted final drive chain	Adjust or renew as necessary.
	Hooked or badly worn sprockets	Renew as a pair.
	Loose rear sprocket	Check sprocket retaining bolts.

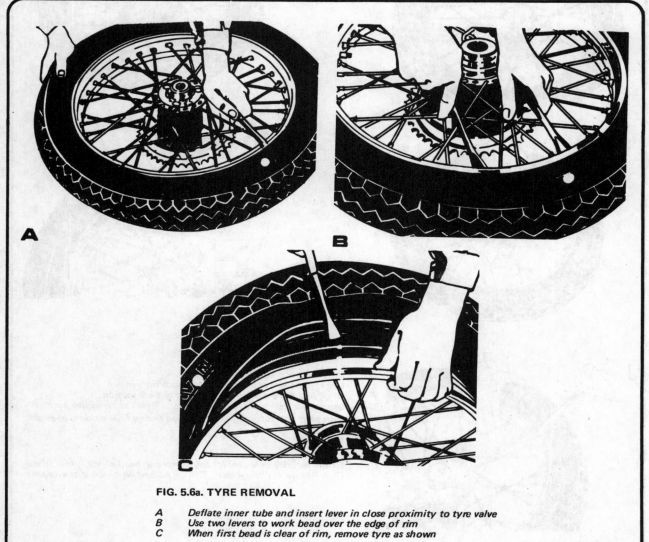

FIG. 5.6a. TYRE REMOVAL

A   *Deflate inner tube and insert lever in close proximity to tyre valve*
B   *Use two levers to work bead over the edge of rim*
C   *When first bead is clear of rim, remove tyre as shown*

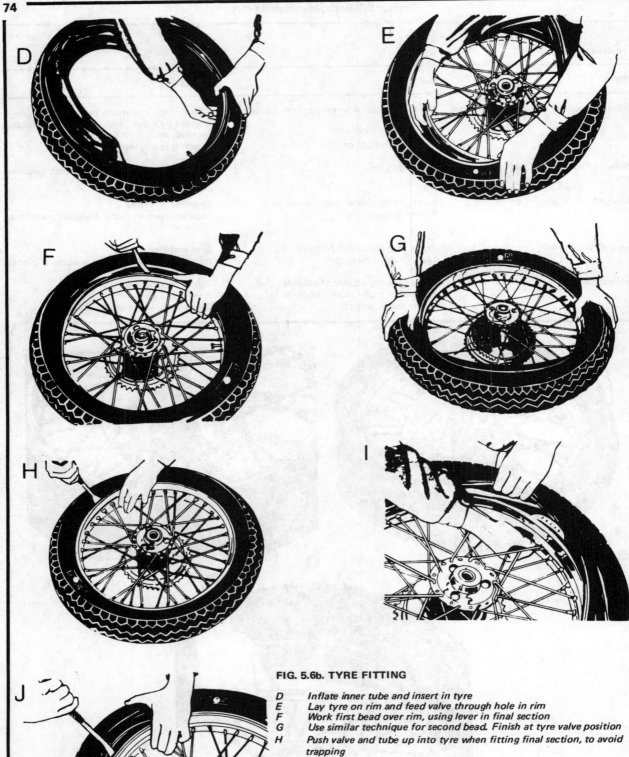

FIG. 5.6b. TYRE FITTING

D	Inflate inner tube and insert in tyre
D E	Lay tyre on rim and feed valve through hole in rim
F	Work first bead over rim, using lever in final section
G	Use similar technique for second bead. Finish at tyre valve position
H	Push valve and tube up into tyre when fitting final section, to avoid trapping

Security bolts

I	Fit the security bolt very loosely when one bead of the tyre is fitted
J	Then fit tyre in normal way. Tighten bolt when tyre is properly seated

Chapter 6 Electrical system

Contents

Specifications

Battery: ... ... Lead acid
    Make ... ... Lucas
    Type PUZ5A 12v
    or 2 x MK9E 6v

Alternator:
    Make ... ... Lucas
    Type ... ... RM19 (pre. 1970)
    RM21 (post 1970)
    RM19ET (Firebird and Hornet)
    Contact breaker ... ... Lucas 4CA or 6CA
    Rectifier ... ... Lucas 2DS 506
    Zener diode type ... ... Lucas ZD 715
    Coils ... ... Lucas 17M12 - 2 off
    Lucas 3ET - 2 off (Firebird and Hornet)
    Ignition capacitor ... ... Lucas 2MC
    Fuse rating ... ... 35 amp

Bulbs:
    Headlamp ... ... 50/40 watt. Prefocus Lucas 12v *
    Pilot ... ... 6 watt 12volt * Lucas 989
    Stop and tail ... ... 6/21 watt 12 volt* Lucas 380
    Speedometer lamp ... ... 2.2 watt 12 volt * Smiths type
    Tachometer lamp ... ... 2.2 watt 12 volt * Smiths type
    Ignition warning lamp ... ... 2 watt 12 volt * Lucas 281
    Main beam warning lamp ... ... 2 watt 12 volt * Lucas 281
    Indicator warning lamp ... ... 2 watt 12 volt * Lucas 281
    Flashing indicators lamps ... ... 21 watt 12 volt * Lucas 382

*Models manufactured before 1966 have a 6 volt system (12 volt equipment optional from 1963 on)

1 General description

An alternating current generator driven from the end of the crankshaft powers the electrical system. The output is converted into direct current by a silicon diode rectifier and supplied to a 12 volt battery, 6 volt, pre 1967 models. An electrical device, known as a Zener diode regulates the charge rate to suit the condition of the battery.

The ignition system, as described in Chapter 5, derives its supply from the rectified current. No emergency start facility is provided because even with a 'flat' battery, the generator output is sufficient to provide the necessary spark.

2 Alternator - checking output

1 The output and performance of the alternator fitted to the BSA unit-construction twins can be checked only with specialised test equipment of the multi-meter type. It is unlikely that the averarage owner will have access to this type of equipment or instruction in its use. In consequence, if the performance is suspect, the alternator and charging circuit should be checked by a qualified auto-electrical expert.

2 Failure of the alternator does not necessarily mean that a replacement is needed. This can however sometimes be most economic through a service exchange scheme. It is possible to replace or rewind the stator coil assembly, for example if the rotor is damaged.

3 If the generator fails to charge, a warning light in the headlamp shell will indicate.

3 Battery - charging procedure and maintenance

1 Whilst the machine is used on the road it is unlikely that the battery will require attention other than routine maintenance because the generator will keep it fully charged. However, if the machine is used for a succession of short journeys only, mainly during the hours of darkness when the lights are in full use, it is possible that the output from the generator may fail to keep pace with the heavy electrical demand, especially if the machine is parked with the lights switched on. Under these circumstances, it will be necessary to remove the battery from time to time to have it charged independently.

2 The battery is located below the dual seat, in a carrier slung between the two parallel frame tubes. It is secured by a strap which when released, will permit the battery to be withdrawn after disconnection of the leads. The battery positive is always earthed.

3 The normal charge rate is 1 amp. A more rapid charge can be given in an emergency, but this should be avoided if possible because it will shorten the life of the battery.

4 When the battery is removed from the machine, remove the cover and clean the battery top. If the terminals are corroded, scrape them clean and cover them with vaseline (not grease) to protect them from further attack. If a vent tube is fitted, make sure it is not obstructed and that it is arranged so that it will not discharge over any parts of the machine.

5 If the machine is laid up for any period of time, the battery should be removed and given a 'refresher' charge every six weeks or so, in order to maintain it in good condition.

6 When two 6 volt batteries are fitted, they must be connected in series with one another. The negative of one battery must go to the wiring harness and the positive of the OTHER to the frame or earth connection. The intermediate connection is made by joining the free negative terminal of one battery to the free positive terminal of the other.

4 Silicon diode rectifier - general

1 The silicon diode rectifier is bolted to a bracket attached to the rear of the battery carrier, beneath the dual seat. Its function is to convert the alternating current from the alternator to direct current which can be used to charge the battery and operate the ignition circuit.

2 The rectifier is deliberately placed in this location so that it is not exposed directly to water or oil and yet has free circulation of air to permit cooling. It should be kept clean and dry; the nuts connecting the rectifier plates should not be disturbed under any circumstances.

3 It is not possible to check whether the rectifier is functioning correctly without the appropriate test equipment. If performance is suspect, a Triumph agent or an auto-electrical expert should be consulted. Note that the rectifier will be destroyed if it is subjected to a reverse flow of current.

4 When tightening the rectifier securing nut, hold the nut at the other end with a spanner. Apart from the fact that the securing stud is sheared very easily, if overtightened, there is risk of the plates twisting and severing their internal connections.

5 Zener diode general

1 The Zener diode is used to regulate the amount of rectified current supplied to charge the battery.

2 In use the Zener diode becomes quite hot as it has to dissipate excess electrical energy in the form of heat and it is customary to fit the diode to a 'heat sink' which will disperse this heat by means of radiating fins. Models before 1971 have the diode mounted on its heat sink beneath the headlamp whereas part 1971 models use the aluminium body of the air cleaner for its mounting point and heat sink.

3 Specialised test equipment is required for testing the Zener diode. If the diode appears to malfunction, a BSA agent or an auto-electrical expert should be consulted. The Zener diode is removed by withdrawing the rubber plug in the forward end of the heat sink, which will expose the mounting nut. Remove the spade connector to the diode first, then unscrew the mounting nut and withdraw the diode, leaving the heat sink in position.

4 Do not make any connection between the diode mounting and the heat sink, otherwise the heat transfer will be affected, resulting in failure of the diode. When replacing the diode, note that the mounting stud is of copper which will shear easily if overstressed.

6 Headlamp - replacing bulbs and adjusting beam light

1 Two types of headlamps are fitted, depending on the year of manufacture of the machine. The pre 1971 models have a headlamp of conventional shape that utilises a reflector unit of the pre-fous type. Post 1970 machines are fitted with a 'short' headlamp, having a recessed back which carries the oil pressure warning, ignition warning and main beam indicator lamps. The reflector unit is of the pre-focus type.

2 To replace either the main bulb or pilot lamp bulb, it is necessary to remove the front of the headlamp in order to gain access to the rear of the reflector unit. Slacken the screw at the top of the headlamp rim and pull the rim complete with reflector unit from the headlamo shell. To remove the contacts from the main bulb holder, press and twist the cover. The main bulb can then be extracted complete with its locating plate and the replacement inserted. The locating plate obviates the need for refocusing, when the bulb is changed. The bulb is locked in position by the contact cover, having a triple bayonet connection offset to ensure that the correct connections are made.

3 The pilot bulb holder is a push fit in the reflector shell. When the holder is withdrawn, the bulb can be released from its bayonet fitting by pressing downward and turning.

4 The main bulb is rated at 50/40, and the pilot lamp bulb at 6 W.

5 After the reflector and headlamp rim have been replaced, beam height can be adjusted by slackening the two bolts which secure the headlamp shell to the forks or, if flashing indicators are fitted, by slackening the locknut around the arms which support the indicator lamps. Adjustments can now be made by tilting the headlamp upward or downward whilst the rider is seated normally. If a pillion passenger is carried, the adjustment made should take this into account because the headlamp will be raised as a result of the increased loading on the rear of the machine.

6 UK Lighting Regulations stipulate that the lighting system should be arranged so that the main beam will not dazzle a person standing in the same horizontal plane as the vehicle at a distance greater than 25 yards from the lamp, whose eye level is not less than 3 feet 6 inches above that plane. It is easy to approximate this setting by placing the machine 25 yards away from a wall, on a level surface, and setting the beam height so

6.1 The interior of the headlamp shell showing colour coded wires and warning light connections

6.2 Press and twist the cover to reveal the headlamp bulb

7.1 The stop/tail light behind its plastic cover

that the beam is concentrated at the same height as the distance from the centre of the headlamp to the ground. The rider must be seated normally during this operation and the pillion passenger, if one is carried regularly.

7 Tail and stop lamp - replacing bulb

1 The combined tail and stop lamp is fitted with a double filament hulb with offset pins to prevent its unintentional reversal in the bulb holder. The lamp unit serves a two-fold purpose; to illuminate the rear of the machine and the rear number plate, and to give visual warning when the rear brake is applied. To gain access to the bulb, remove the two screws securing below it. The bulb is released by pressing inward with a twisting action; it is rated at 6/21W.
2 The stop lamp is actuated by a switch bolted to the rear chainguard. The switch is connected to the brake rod by a spring attached to a clamp fitting around the rod; the point at which the switch operates is governed by the position of the clamp in relation to the rod. If the clamp is moved forward, the stop lamp wll indicate earlier and vice versa. The switch does not require attention other than the occasional drop of thin oil.

8 Speedometer and tachometer bulbs - replacement

The speedometer and tachometer heads are each fitted with a bulb to illuminate the dial where the headlamp is switched on. The bulb holders are a push fit into the bottom of each instrument case and carry a 2.2W bulb which has a threaded body.

9 Ignition warning, oil pressure warning, main beam and flashing indicator warning bulbs - replacement

1 The combination of lamps fitted varies according to the model and year of manufacture. There is not necessarily four separate forms of warning fitted to each machine.
2 The bulb holders are a push fit into either the top of the back of the headlamp shell, depending on the type of headlamp fitted. Each of the bulbs is rated at 2W.

10 Flashing indicator lamps

1 Late models have flashing direction indicator lamps attached to the front and rear of the machine. They are operated by a thumb switch on the left hand end of the handlebars. An indicator lamp built into the rear of the headlamp shell will flash in unison with the lights, provided the front and rear lights are operating correctly.
2 The bulbs are fitted by removing the plastic lens covers, held in position by two screws. The bulbs are of the bayonet type and must be pressed and turned to release or fit. Each bulb is rated at 21W.

11 Flasher unit - location and replacement

1 The flasher unit is located beneath the dual seat, along with the other electrical equipment. It seldom gives trouble unless it is subjected to a heavy blow which will disturb its senstive action.
2 It is not possible to renovate a manfunctioning flasher unit. If the bulbs are in working order and will give only a single flash when the handlebar switch is operated, the flasher unit should be suspected and, if necessary, renewed.

12 Headlamp dip switch

1 The headlamp dip switch forms part of the switch unit fitted

to the right hand side of the handlebars on all late models. Earlier models have a separate dip switch mounted on the left hand side of the handlebars which also contain the horn push.

2   If the dipswitch malfunctions, the switch unit must be renewed since it is seldom practicable to effect a satisfactory repair.

13 Horn push and horn - adjustment

1   The horn push on late models forms part of the switch unit at the right hand end of the handlebars. On earlier models, it is combined with the separate dip switch.

2   The horn is secured below the nose of the petrol tank, facing in a forward direction. It is provided with adjustment in the form of a serrated screw inset into the back of the horn body.

3   To adjust the horn, turn the screw anti-clockwise until the horn just fails to sound, then back it off about one-quarter turn. Adjustment is needed only very occasionally, to compensate for wear of the internal moving parts.

14 Fuse - location and replacement

1   A fuse is incorporated in the brown/blue coloured lead from the negative terminal of the battery. It is housed within a quickly detachable shell and protects the electrical equipment from accidental damage if a short circuit should occur.

2   If the electrical system will not operate, a blown fuse should be suspected, but before the fuse is renewed, the electrical system should be inspected to trace the reason for the failure of the fuse. If this precaution is not observed, the replacement fuse may blow too.

3   The fuse is rated at 35 amps and at least one spare should always be carried. In an extreme emergency, when the cause of the failure has been rectified and if no spare is available, a get-you-home repair can be effected by wrapping silver paper around the blown fuse and re-inserting it in the fuse holder. It must be stressed that this in only an emergency measure and the 'bastard' fuse should be replaced at the earliest possible opportunity. It affords no protection whatsoever to the electrical circuit when bridged in this fashion.

15 Ignition switch

1   The ignition switch is fitted to the left hand top cover of the forks, or in the left hand or right hand cover surrounding the rear portion of the frame, depending on the model and year of manufacture.

2   It is retained by a locknut or a locking ring which, when unscrewed, will free the switch.

16 Headlamp switch

1   A two or three position headlamp switch is fitted to the headlamp shell to operate the pilot and main headlamp bulbs. Machines fitted with the two-position rotary switch must have the ignition switch in position 4 before the headlamp will operate.

2   Late models have an additional headlamp flasher in the form of a push button embodied in the switch assembly on the right hand end of the handlebars.

17 Ammeter

1   All models with the full headlamp shell have an ammeter inset into the top of the headlamp shell to show the amount of charge from the generator.

2   When an ammeter is not fitted, as in the case of the late models with the 'short' headlamp shell, an ignition warning light is fitted. This light will not extinguish after the engine is started, if the generator has failed.

18 Capacitor ignition

1   A capacitor is built into the ignition circuit so that a machine without lights (and therefore without a battery) or one on which the battery has failed, can be started and run normally. If lights are fitted, the machine can be used during the hours of darkness, since the lighting equipment will function correctly immediately the engine starts, even if the battery is removed.

2   The capacitor is fitted into a coil spring to protect it from vibration. It is normally mounted with its terminals pointing downward from a convenient point underneath the dual seat.

3   Before running a machine on the capacitor system with the battery disconnected, it is necessary to tape up the battery negative so that it cannot reconnect accidentally and short circuit. If this occurs, the capacitor will be ruined. A convenient means of isolating the battery is to remove the fuse.

19 Wiring - layout and examination

1   The cables of the wiring harness are colour-coded and will correspond with the accompanying wiring diagrams.

2   Visual inspection will show whether any breaks or frayed outer coverings are giving rise to short circuits which will cause the main fuse to blow. Another source of trouble is the snap connectors and spade terminals, which may make a poor connection if they are not pushed home fully.

3   Intermittent short circuits can sometimes be traced to a chafed wire passing through, or close to, a metal component, such as a frame member. Avoid tight bends in the cables or situations where the cables can be trapped or stretched especially in the vicinity of the handlebars or steering head.

20 Front brake stop lamp switch

1   In order to comply with traffic requirements in certain overseas countries, a stop lamp switch is now incorporated in the front brake cable so that the rear stop lamp is illuminated when the front brake is applied.

2   There is no means of adjustment for the front brake stop lamp switch. If the switch malfunctions, the front brake cable must be replaced.

21 Fault diagnosis

Symptom	Cause	Remedy
Complete electrical failure	Blown fuse	Check wiring and electrical components for short circuit before fitting new 15 amp fuse.
	Isolated battery	Check battery connections, also whether connections show signs of corrosion.
Dim lights, horn and starter inoperative	Discharged battery	Recharge battery with battery charger. Check whether generator is giving correct output.
Constantly blowing bulbs	Vibration, poor earth connection	Check security of bulb holders. Check earth return connections.

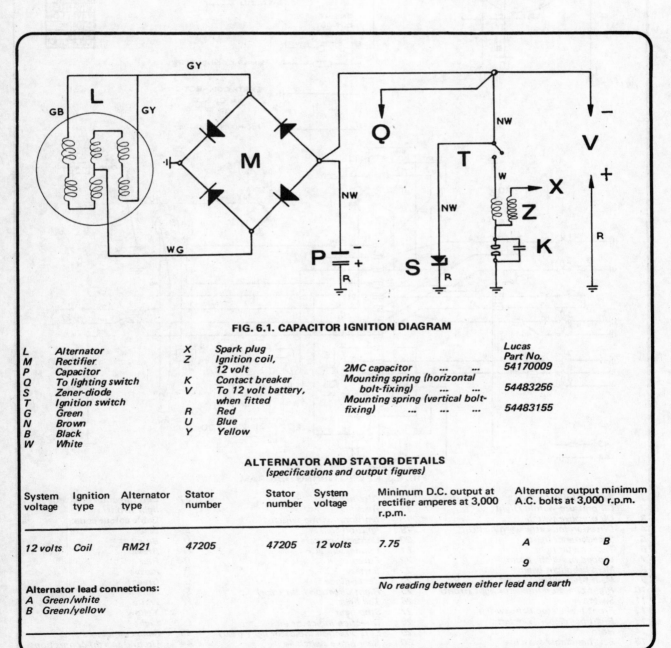

FIG. 6.1. CAPACITOR IGNITION DIAGRAM

					Lucas Part No.
L	Alternator	X	Spark plug		
M	Rectifier	Z	Ignition coil, 12 volt	2MC capacitor	54170009
P	Capacitor	K	Contact breaker	Mounting spring (horizontal bolt-fixing)	54483256
Q	To lighting switch	V	To 12 volt battery, when fitted	Mounting spring (vertical bolt-fixing)	54483155
S	Zener-diode	R	Red		
T	Ignition switch	U	Blue		
G	Green	Y	Yellow		
N	Brown				
B	Black				
W	White				

ALTERNATOR AND STATOR DETAILS
(specifications and output figures)

System voltage	Ignition type	Alternator type	Stator number	Stator number	System voltage	Minimum D.C. output at rectifier amperes at 3,000 r.p.m.	Alternator output minimum A.C. bolts at 3,000 r.p.m.	
12 volts	Coil	RM21	47205	47205	12 volts	7.75	A	B
							9	0

No reading between either lead and earth

Alternator lead connections:
A Green/white
B Green/yellow

80

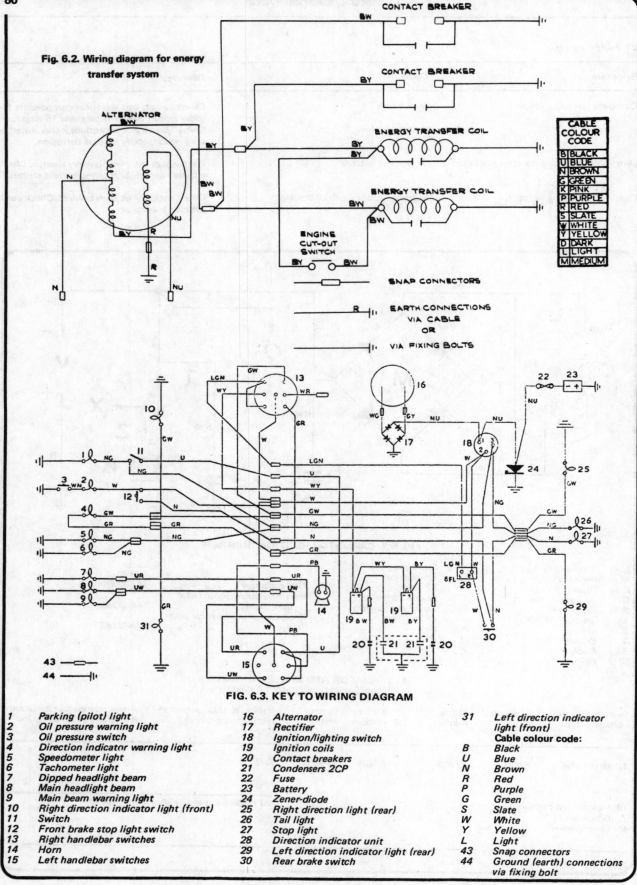

Fig. 6.2. Wiring diagram for energy transfer system

FIG. 6.3. KEY TO WIRING DIAGRAM

1	Parking (pilot) light	16	Alternator	31	Left direction indicator light (front)
2	Oil pressure warning light	17	Rectifier		**Cable colour code:**
3	Oil pressure switch	18	Ignition/lighting switch	B	Black
4	Direction indicator warning light	19	Ignition coils	U	Blue
5	Speedometer light	20	Contact breakers	N	Brown
6	Tachometer light	21	Condensers 2CP	R	Red
7	Dipped headlight beam	22	Fuse	P	Purple
8	Main headlight beam	23	Battery	G	Green
9	Main beam warning light	24	Zener-diode	S	Slate
10	Right direction indicator light (front)	25	Right direction light (rear)	W	White
11	Switch	26	Tail light	Y	Yellow
12	Front brake stop light switch	27	Stop light	L	Light
13	Right handlebar switches	28	Direction indicator unit	43	Snap connectors
14	Horn	29	Left direction indicator light (rear)	44	Ground (earth) connections via fixing bolt
15	Left handlebar switches	30	Rear brake switch		

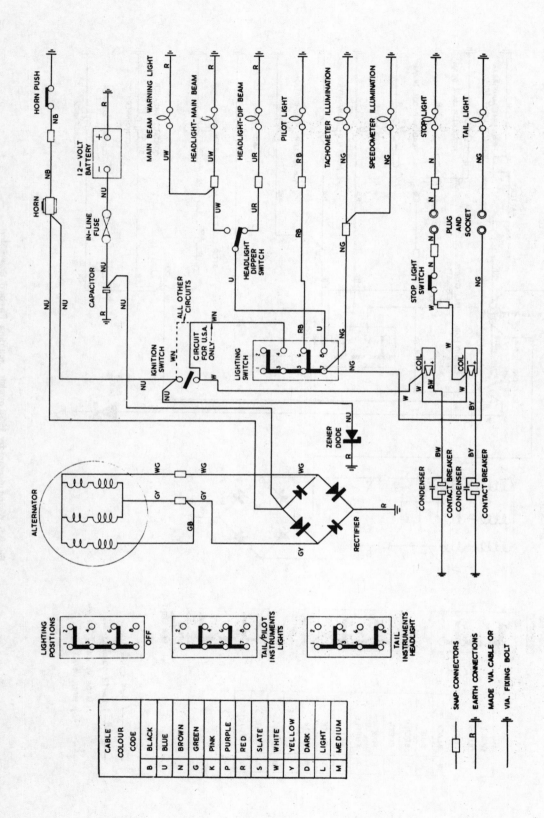

Fig. 6.4. Wiring diagram for 'Firebird' Scrambler

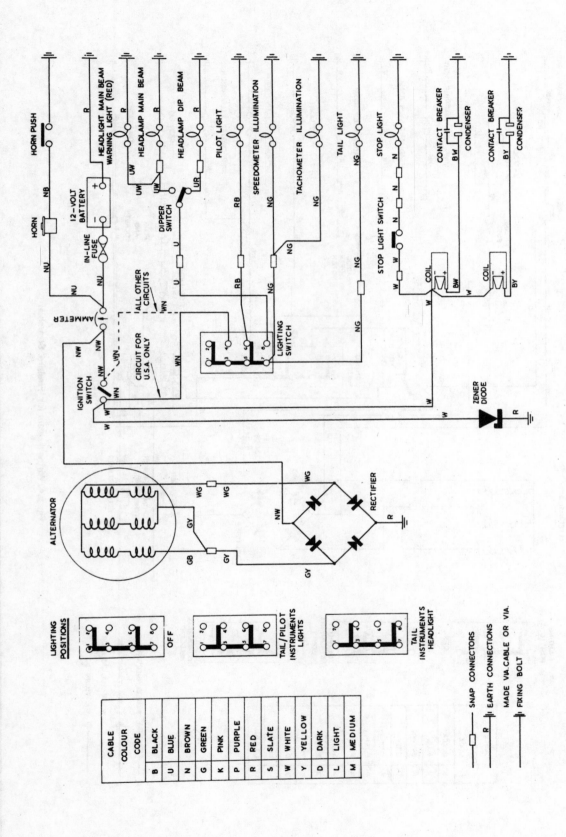

Fig. 6.5. Wiring diagram for 1968 models, excluding 'Firebird' Scrambler

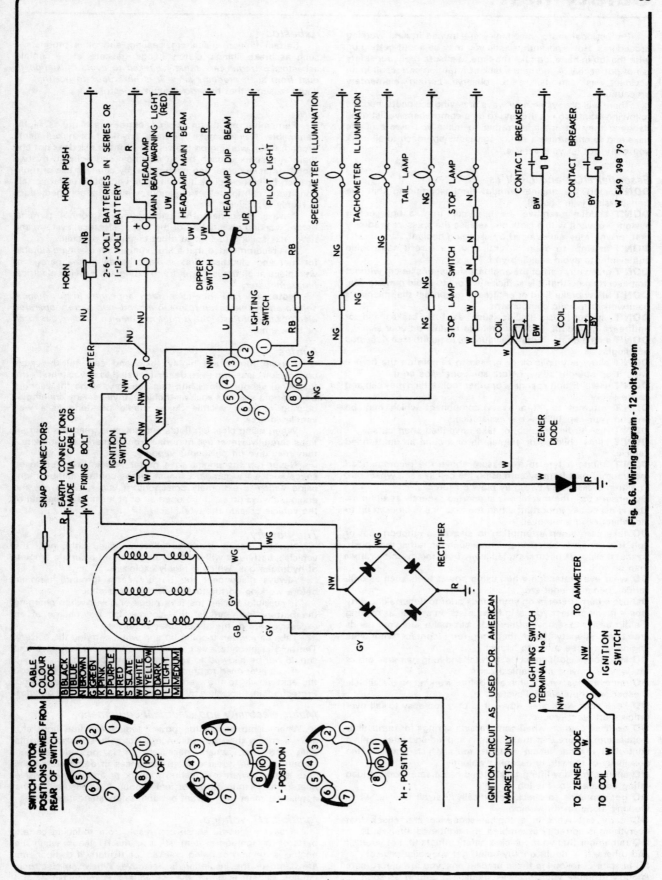

Fig. 6.6. Wiring diagram - 12 volt system

Safety first!

Professional motor mechanics are trained in safe working procedures. However enthusiastic you may be about getting on with the job in hand, do take the time to ensure that your safety is not put at risk. A moment's lack of attention can result in an accident, as can failure to observe certain elementary precautions.

There will always be new ways of having accidents, and the following points do not pretend to be a comprehensive list of all dangers; they are intended rather to make you aware of the risks and to encourage a safety-conscious approach to all work you carry out on your vehicle.

Essential DOs and DON'Ts

DON'T start the engine without first ascertaining that the transmission is in neutral.

DON'T suddenly remove the filler cap from a hot cooling system – cover it with a cloth and release the pressure gradually first, or you may get scalded by escaping coolant.

DON'T attempt to drain oil until you are sure it has cooled sufficiently to avoid scalding you.

DON'T grasp any part of the engine, exhaust or silencer without first ascertaining that it is sufficiently cool to avoid burning you.

DON'T allow brake fluid or antifreeze to contact the machine's paintwork or plastic components.

DON'T syphon toxic liquids such as fuel, brake fluid or antifreeze by mouth, or allow them to remain on your skin.

DON'T inhale dust – it may be injurious to health (see *Asbestos* heading).

DON'T allow any spilt oil or grease to remain on the floor – wipe it up straight away, before someone slips on it.

DON'T use ill-fitting spanners or other tools which may slip and cause injury.

DON'T attempt to lift a heavy component which may be beyond your capability – get assistance.

DON'T rush to finish a job, or take unverified short cuts.

DON'T allow children or animals in or around an unattended vehicle.

DON'T inflate a tyre to a pressure above the recommended maximum. Apart from overstressing the carcase and wheel rim, in extreme cases the tyre may blow off forcibly.

DO ensure that the machine is supported securely at all times. This is especially important when the machine is blocked up to aid wheel or fork removal.

DO take care when attempting to slacken a stubborn nut or bolt. It is generally better to pull on a spanner, rather than push, so that if slippage occurs you fall away from the machine rather than on to it.

DO wear eye protection when using power tools such as drill, sander, bench grinder etc.

DO use a barrier cream on your hands prior to undertaking dirty jobs – it will protect your skin from infection as well as making the dirt easier to remove afterwards; but make sure your hands aren't left slippery. Note that long-term contact with used engine oil can be a health hazard.

DO keep loose clothing (cuffs, tie etc) and long hair well out of the way of moving mechanical parts.

DO remove rings, wristwatch etc, before working on the vehicle – especially the electrical system.

DO keep your work area tidy – it is only too easy to fall over articles left lying around.

DO exercise caution when compressing springs for removal or installation. Ensure that the tension is applied and released in a controlled manner, using suitable tools which preclude the possibility of the spring escaping violently.

DO ensure that any lifting tackle used has a safe working load rating adequate for the job.

DO get someone to check periodically that all is well, when working alone on the vehicle.

DO carry out work in a logical sequence and check that everything is correctly assembled and tightened afterwards.

DO remember that your vehicle's safety affects that of yourself and others. If in doubt on any point, get specialist advice.

IF, in spite of following these precautions, you are unfortunate enough to injure yourself, seek medical attention as soon as possible.

Asbestos

Certain friction, insulating, sealing, and other products – such as brake linings, clutch linings, gaskets, etc – contain asbestos. *Extreme care must be taken to avoid inhalation of dust from such products since it is hazardous to health*. If in doubt, assume that they *do* contain asbestos.

Fire

Remember at all times that petrol (gasoline) is highly flammable. Never smoke, or have any kind of naked flame around, when working on the vehicle. But the risk does not end there – a spark caused by an electrical short-circuit, by two metal surfaces contacting each other, by careless use of tools, or even by static electricity built up in your body under certain conditions, can ignite petrol vapour, which in a confined space is highly explosive.

Always disconnect the battery earth (ground) terminal before working on any part of the fuel or electrical system, and never risk spilling fuel on to a hot engine or exhaust.

It is recommended that a fire extinguisher of a type suitable for fuel and electrical fires is kept handy in the garage or workplace at all times. Never try to extinguish a fuel or electrical fire with water.

Note: *Any reference to a 'torch' appearing in this manual should always be taken to mean a hand-held battery-operated electric lamp or flashlight. It does **not** mean a welding/gas torch or blowlamp.*

Fumes

Certain fumes are highly toxic and can quickly cause unconsciousness and even death if inhaled to any extent. Petrol (gasoline) vapour comes into this category, as do the vapours from certain solvents such as trichloroethylene. Any draining or pouring of such volatile fluids should be done in a well ventilated area.

When using cleaning fluids and solvents, read the instructions carefully. Never use materials from unmarked containers – they may give off poisonous vapours.

Never run the engine of a motor vehicle in an enclosed space such as a garage. Exhaust fumes contain carbon monoxide which is extremely poisonous; if you need to run the engine, always do so in the open air or at least have the rear of the vehicle outside the workplace.

The battery

Never cause a spark, or allow a naked light, near the vehicle's battery. It will normally be giving off a certain amount of hydrogen gas, which is highly explosive.

Always disconnect the battery earth (ground) terminal before working on the fuel or electrical systems.

If possible, loosen the filler plugs or cover when charging the battery from an external source. Do not charge at an excessive rate or the battery may burst.

Take care when topping up and when carrying the battery. The acid electrolyte, even when diluted, is very corrosive and should not be allowed to contact the eyes or skin.

If you ever need to prepare electrolyte yourself, always add the acid slowly to the water, and never the other way round. Protect against splashes by wearing rubber gloves and goggles.

Mains electricity and electrical equipment

When using an electric power tool, inspection light etc, always ensure that the appliance is correctly connected to its plug and that, where necessary, it is properly earthed (grounded). Do not use such appliances in damp conditions and, again, beware of creating a spark or applying excessive heat in the vicinity of fuel or fuel vapour. Also ensure that the appliances meet the relevant national safety standards.

Ignition HT voltage

A severe electric shock can result from touching certain parts of the ignition system, such as the HT leads, when the engine is running or being cranked, particularly if components are damp or the insulation is defective. Where an electronic ignition system is fitted, the HT voltage is much higher and could prove fatal.

Metric conversion tables

Inches	Decimals	Millimetres	Millimetres to Inches		Inches to Millimetres	
			mm	Inches	Inches	mm
1/64	0.015625	0.3969	0.01	0.00039	0.001	0.0254
1/32	0.03125	0.7937	0.02	0.00079	0.002	0.0508
3/64	0.046875	1.1906	0.03	0.00118	0.003	0.0762
1/16	0.0625	1.5875	0.04	0.00157	0.004	0.1016
5/64	0.078125	1.9844	0.05	0.00197	0.005	0.1270
3/32	0.09375	2.3812	0.06	0.00236	0.006	0.1524
7/64	0.109375	2.7781	0.07	0.00276	0.007	0.1778
1/8	0.125	3.1750	0.08	0.00315	0.008	0.2032
9/64	0.140625	3.5719	0.09	0.00354	0.009	0.2286
5/32	0.15625	3.9687	0.1	0.00394	0.01	0.254
11/64	0.171875	4.3656	0.2	0.00787	0.02	0.508
3/16	0.1875	4.7625	0.3	0.01181	0.03	0.762
13/64	0.203125	5.1594	0.4	0.01575	0.04	1.016
7/32	0.21875	5.5562	0.5	0.01969	0.05	1.270
15/64	0.234375	5.9531	0.6	0.02362	0.06	1.524
1/4	0.25	6.3500	0.7	0.02756	0.07	1.778
17/64	0.265625	6.7469	0.8	0.03150	0.08	2.032
9/32	0.28125	7.1437	0.9	0.03543	0.09	2.286
19/64	0.296875	7.5406	1	0.03937	0.1	2.54
5/16	0.3125	7.9375	2	0.07874	0.2	5.08
21/64	0.328125	8.3344	3	0.11811	0.3	7.62
11/32	0.34375	8.7312	4	0.15748	0.4	10.16
23/64	0.359375	9.1281	5	0.19685	0.5	12.70
3/8	0.375	9.5250	6	0.23622	0.6	15.24
25/64	0.390625	9.9219	7	0.27559	0.7	17.78
13/32	0.40625	10.3187	8	0.31496	0.8	20.32
27/64	0.421875	10.7156	9	0.35433	0.9	22.86
7/16	0.4375	11.1125	10	0.39370	1	25.4
29/64	0.453125	11.5094	11	0.43307	2	50.8
15/32	0.46875	11.9062	12	0.47244	3	76.2
31/64	0.484375	12.3031	13	0.51181	4	101.6
1/2	0.5	12.7000	14	0.55118	5	127.0
33/64	0.515625	13.0969	15	0.59055	6	152.4
17/32	0.53125	13.4937	16	0.62992	7	177.8
35/64	0.546875	13.8906	17	0.66929	8	203.2
9/16	0.5625	14.2875	18	0.70866	9	228.6
37/64	0.578125	14.6844	19	0.74803	10	254.0
19/32	0.59375	15.0812	20	0.78740	11	279.4
39/64	0.609375	15.4781	21	0.82677	12	304.8
5/8	0.625	15.8750	22	0.86614	13	330.2
41/64	0.640625	16.2719	23	0.90551	14	355.6
21/32	0.65625	16.6687	24	0.94488	15	381.0
43/64	0.671875	17.0656	25	0.98425	16	406.4
11/16	0.6875	17.4625	26	1.02362	17	431.8
45/64	0.703125	17.8594	27	1.06299	18	457.2
23/32	0.71875	18.2562	28	1.10236	19	482.6
47/64	0.734375	18.6531	29	1.14173	20	508.0
3/4	0.75	19.0500	30	1.18110	21	533.4
49/64	0.765625	19.4469	31	1.22047	22	558.8
25/32	0.78125	19.8437	32	1.25984	23	584.2
51/64	0.796875	20.2406	33	1.29921	24	609.6
13/16	0.8125	20.6375	34	1.33858	25	635.0
53/64	0.828125	21.0344	35	1.37795	26	660.4
27/32	0.84375	21.4312	36	1.41732	27	685.8
55/64	0.859375	21.8281	37	1.4567	28	711.2
7/8	0.875	22.2250	38	1.4961	29	736.6
57/64	0.890625	22.6219	39	1.5354	30	762.0
29/32	0.90625	23.0187	40	1.5748	31	787.4
59/64	0.921875	23.4156	41	1.6142	32	812.8
15/16	0.9375	23.8125	42	1.6535	33	838.2
61/64	0.953125	24.2094	43	1.6929	34	863.6
31/32	0.96875	24.6062	44	1.7323	35	889.0
63/64	0.984375	25.0031	45	1.7717	36	914.4

English/American terminology

Because this book has been written in England, British English component names, phrases and spellings have been used throughout. American English usage is quite often different and whereas normally no confusion should occur, a list of equivalent terminology is given below.

English	American	English	American
Air filter	Air cleaner	Number plate	License plate
Alignment (headlamp)	Aim	Output or layshaft	Countershaft
Allen screw/key	Socket screw/wrench	Panniers	Side cases
Anticlockwise	Counterclockwise	Paraffin	Kerosene
Bottom/top gear	Low/high gear	Petrol	Gasoline
Bottom/top yoke	Bottom/top triple clamp	Petrol/fuel tank	Gas tank
Bush	Bushing	Pinking	Pinging
Carburettor	Carburetor	Rear suspension unit	Rear shock absorber
Catch	Latch	Rocker cover	Valve cover
Circlip	Snap ring	Selector	Shifter
Clutch drum	Clutch housing	Self-locking pliers	Vise-grips
Dip switch	Dimmer switch	Side or parking lamp	Parking or auxiliary light
Disulphide	Disulfide	Side or prop stand	Kick stand
Dynamo	DC generator	Silencer	Muffler
Earth	Ground	Spanner	Wrench
End float	End play	Split pin	Cotter pin
Engineer's blue	Machinist's dye	Stanchion	Tube
Exhaust pipe	Header	Sulphuric	Sulfuric
Fault diagnosis	Trouble shooting	Sump	Oil pan
Float chamber	Float bowl	Swinging arm	Swingarm
Footrest	Footpeg	Tab washer	Lock washer
Fuel/petrol tap	Petcock	Top box	Trunk
Gaiter	Boot	Torch	Flashlight
Gearbox	Transmission	Two/four stroke	Two/four cycle
Gearchange	Shift	Tyre	Tire
Gudgeon pin	Wrist/piston pin	Valve collar	Valve retainer
Indicator	Turn signal	Valve collets	Valve cotters
Inlet	Intake	Vice	Vise
Input shaft or mainshaft	Mainshaft	Wheel spindle	Axle
Kickstart	Kickstarter	White spirit	Stoddard solvent
Lower leg	Slider	Windscreen	Windshield
Mudguard	Fender		

Index